Research Reports ESPRIT

Project 1072 · DIAMOND · Vol. 1

Edited in cooperation with
the Commission of the European Communities

Ch. Ullrich   J. Wolff von Gudenberg (Eds.)

Accurate Numerical Algorithms

A Collection of Research Papers

Springer-Verlag
Berlin Heidelberg New York
London Paris Tokyo Hong Kong

Editors

Ch. Ullrich
Universität Basel, Institut für Informatik
Mittlere Straße 142, CH–4056 Basel, Switzerland

J. Wolff von Gudenberg
Universität Karlsruhe, Institut für Angewandte Mathematik
Kaiserstraße 12, D–7500 Karlsruhe 1, FRG

ESPRIT-Project 1072 "Development and Integration of Accurate Mathematical
Operations in Numerical Data Processing (DIAMOND)" belongs to the
Subprogramme Software Technology (ST) of ESPRIT, the European Strategic
Programme for Research and Development in Information Technology supported
by the European Communities.

The aim of this project is the development of methods and tools allowing accurate floating point arithmetic on computers, based on a mathematical theory of
computer arithmetic in which all operations are defined by so-called
semimorphisms. Such systematic theory of computer arithmetic aims at
performing the basic arithmetic operations to maximum accuracy and providing
sufficient control over the rounding process so as to ensure reliable error
bounds. This project will pursue different approaches: embedding of convenient
arithmetic notations into Ada and Pascal; AI techniques for formula
transformation and symbolic manipulation; construction of a methodological
framework and a knowledge base for numerical programming.

The project results are expected to have a strong impact on accuracy, safety,
reliability and economy of numerical programming, a field of interest for many
scientific and industrial domains.

Participating Organisations:
Siemens/FRG, CWI/The Netherlands, University of Karlsruhe/FRG, NAG/UK,
University of Bath/UK

ISBN-13:978-3-540-51477-0   e-ISBN-13:978-3-642-83874-3
DOI: 10.1007/978-3-642-83874-3

2145/3140 – 543210 – Printed on acid-free paper

Preface

The ESPRIT Project 1072, DIAMOND (Development and Integration of Accurate
Mathematical Operations in Numerical Data-Processing) was carried out from
January 1986 through April 1989 by the five partners Siemens München (prime
contractor), CWI Amsterdam, University of Karlsruhe (Institut für Angewandte Mathematik), NAG Oxford and University of Bath (subcontractor to
NAG). The technical work was divided into three main work packages with one
additional work package for miscellaneous topics.

The major goals of this project, according to its title, were to develop a
set of accurate numerical algorithms (work package 3) and to provide tools
to support their implementation by means of an embedding of accurate
arithmetic into programming languages (work package 1) and by transformation techniques which either improve the accuracy of expression evaluation
or detect and eliminate presumable deficiencies in accuracy in existing
programs (work package 2). A great variety of working papers describing and
discussing the results of these work packages have been written during the
collaboration of the project.

This book (*Accurate Numerical Algorithms*) mainly summarizes the results of
work package 3, carried through by the two partners Karlsruhe University
and NAG Oxford under the leadership of the editors and Dr. G.S. Hodgson.
Another book (*Improving Floating-Point Programming, J. Wiley, to appear
1989*), edited by P.J.L. Wallis, which is one of the DIAMOND project's final
deliverables, concentrates more on the fundamental tools.

Acknowledgements

The work of Karlsruhe University has been supported by the Ministerium für
Wirtschaft, Mittelstand und Technologie, Baden-Württemberg and the Ministerium für Wissenschaft und Kunst, Baden-Württemberg.
We are grateful for the help of all partners including the CEC for their
support in publishing this book and to Springer-Verlag for publication. We
would especially like to thank Miss Susan Brown who helped us to improve
the English of the text.

Karlsruhe, May 1989   Ch. Ullrich
                                                    J. Wolff von Gudenberg

Table of Contents

Highly Accurate Numerical Algorithms
Revised version of DIAMOND Deliverable D3-6

Günter Schumacher
Jürgen Wolff von Gudenberg
Universität Karlsruhe

0. Introduction

The objective of work package 3, solutions to selected numerical problems, has been stated in the Technical Annex as follows:
"This work package intends the development of algorithms which compute verified solutions of numerical problems with high or maximum accuracy, to be written in PASCAL-SC and the language embeddings in Ada achieved in work package 1. Simultaneously, those algorithms would show the existence and uniqueness of the solution.

Program packages are so far available for some standard problems such as real linear systems of equations, inversion of real matrices, real eigen-values and eigenvector problems, real zeros of real polynomials and evaluation of arithmetic real expressions.
Similar algorithms for the solution of complex linear systems of equations, inversion of complex matrices and corresponding complex eigenvalue and eigenvector problems and for the computation of single complex zeros and multiple complex zeros (if possible) of real and complex polynomials have to be developed and implemented. Further investigations will be made in the area of sparse matrix problems, the solution of nonlinear equations, and quadrature.

There will be a requirement for accurate formula evaluation during program execution."

In the following paper we summarize the results of this work package. We have investigated all of the selected problems. Algorithms for a sharp inclusion of the true solution have been developed even for crucial numerical cases.

The following problems have been studied and solved:
- complex linear systems,
- linear systems with band-shaped and general sparse matrices,
- complex eigenvalues and eigenvectors,
- eigenvalues of real symmetric matrices,
- evaluation of complex polynomials,
- simple zeros of complex polynomials,
- inclusion of all (also multiple) zeros of complex polynomials,
- quadrature,
- non-linear systems.

After an introduction of the design methodology of the inclusion algorithms which are called E-methods we present an application of Brouwer's fixed-point theorem as the mathematical background of some algorithms in chapter 2. Chapter 3 describes eigenvalue problems and chapter 4 the application of the Schur-Cohn algorithm using Cauchy indices and the principle of argument. Chapter 5 is devoted to the solution of linear systems with band-shaped or general sparse matrices, whereas chapter 6 shows the application of interval Taylor expansion in numerical quadrature. Finally, chapter 7 deals with nonlinear systems of equations. In this paper no attempt for mathematical completeness was made. It should give the reader a rough but nevertheless introductory overview of those methods.

We do not comment on the solution of complex linear systems and the treatment of complex polynomials which have been described in [12]. All algorithms have been implemented in PASCAL-SC [6] or in Ada. They all use interval arithmetic [1], [24] and accurate arithmetic features in the sense of the Karlsruhe definition [21].
Introduction to Interval Arithmetic [31], the Karlsruhe Accurate Arithmetic Approach [35] as well as its embedding into PASCAL-SC [34] or Ada [20] may be found in the book "Improving Floating-Point Programming" [32].
The PASCAL-SC programs are comprised in an interactive demonstration package whose output for some selected samples is displayed in chapter 8.

1. Design of E-methods

The implemented algorithms are called "E-methods", since they prove the Existence of a result and compute an Enclosure for it. In most cases uniqueness is guaranteed [17]. These algorithms always deliver guaranteed results, such that the true result is enclosed into sharp bounds.

The most obvious way to obtain an inclusion of the true result would be to apply a direct method where all operations are replaced by their corresponding interval operation. This simplistic way may lead to unrealisticly pessimistic bounds. It is this behaviour which has discredited interval arithmetic in the past. For particular problems the method works well, however, as is shown by the implementation of the Schur-Cohn algorithm by means of circular complex interval arithmetic. Since this simplistic way has only a limited range of application other methods have to be developed.

These E-methods are also called "self-validating algorithms", since they automatically verify the result of a numerical computation. This verification is achieved by checking all suppositions of a mathematical theorem during the computation. This is only possible if the used computer arithmetic is faithful in the sense that all unavoidable errors due to roundoff can be described by mathematical error estimates.

The methods themselves usually start with an approximation of the result which can be obtained by conventional numerical methods. After that, they try to apply a mathematical theorem, mostly a fixed-point theorem on the computer. As already mentioned, all errors including round-off errors have to be considered. Therefore, interval arithmetic is mandatory.

In some cases, an iterative refinement technique using higher precision arithmetic is applied to achieve as much accuracy as is needed. Since the exact solution is in general neither computable nor representable on computers, the result is always an enclosing interval representing a lower and upper bound of the solution. The enclosure of the result may not be obtained in all cases, but the algorithms detect if this ever happens and output an appropriate message. Thus, the algorithms are safe in the sense that wrong results will never be produced. The success of the whole process may depend on the accuracy of the approximation. Therefore, well-known approximation algorithms are used and they are improved by the use of highly accurate arithmetic.

In this context, accurate expression evaluation has been studied in work package 2 of the DIAMOND project [9], [10]. It can be shown that it often suffices to evaluate scalar product expressions optimally [21].

In the following chapter we describe E-methods using Brouwer's fixed-point theorem.

In the self-validating algorithm for quadrature (see chapter 6) interval arithmetic is not only applied to handle round-off errors, but also to estimate the approximation error and to control the way of the computation. The partioning of the subintervals is adaptively performed by subdividing the interval whose result has the largest width. In this case the term "interval analysis" may be more appropriate.

2. Application of Brouwer's fixed-point Theorem

A simple consequence of Brouwer's fixed-point Theorem is the following [15]:

Theorem 1:
Let X be a nonempty, convex and compact subset of a finite dimensional linear normed space M. Let f: X → M be a continuous function and F: $\mathbb{P}$X → $\mathbb{P}$M with f(x) ∈ F(A) for all x ∈ A and A ∈ $\mathbb{P}$X. If

$$F(X) \subseteq X \qquad\qquad (1)$$

then the equation f(x) = x has at least one solution in X.

In our application M will be the space of real or complex vectors, respectively, and X will be a corresponding interval vector. An appropriate function f has to be found for a given problem in order to apply the theorem. F is chosen to be the interval extension of f which is obtained by replacing all real operations in f by corresponding interval operations.

Having found an f we have to construct an appropriate interval vector X and check whether $F(X) \subseteq X$ holds. This condition can be verified on a computer by interval arithmetic. Let $\langle\!\langle F \rangle\!\rangle$ denote the computed interval extension of f. Then

$$f(x) \in F(X) \subseteq \langle\!\langle F \rangle\!\rangle (X)$$

always holds. Therefore $\langle\!\langle F \rangle\!\rangle (X) \subseteq X$ implies $F(X) \subseteq X$, and our theorem can be applied.

As an example for the process of finding such an appropriate fixed-point problem we consider systems of algebraic equations.

$$g(x) = 0,$$

where $g: D \subseteq \mathbb{R}^n \to \mathbb{R}^n$ or $g: D \subseteq \mathbb{C}^n \to \mathbb{C}^n$, respectively.
This problem can be transformed to a fixed-point problem by defining

$$x = f(x) := x - R \cdot g(x) \qquad (2)$$

with a nonsingular matrix R.
For a linear system $g(x) := A \cdot x - b$ we have

$$f(x) = x - R(Ax - b) = Rb + (I - RA)x$$

Theorem 2
Let $F(X) := R \cdot b + (I - RA)X$ be the interval extension of $f(x)$. If

$$F(X) \subsetneq X$$

holds, then the equation $g(x) = 0$ and thus the linear system $Ax = b$ has one and only one solution $\hat{x}$ in X and $\hat{x} \in F^k(x)$, $k \geq 0$.

Here $\subsetneq$ denotes the proper inclusion, i.e.

$$A \subsetneq B \iff A \subseteq B \wedge A \neq B$$

A proof of Theorem 2 may be found in [29]. The Theorem can easily be extended to nonlinear systems.

In the following $\underline{\cup}$ denotes the convex hull of two intervals, i.e. the smallest enclosing interval.

Theorem 3:
Let $\tilde{x} \in \mathbb{R}^n$, $X \in I\mathbb{R}^n$, $g: \tilde{x} \underline{\cup} X \to \mathbb{R}^n$, and suppose there exists an interval matrix $L(\tilde{x} \underline{\cup} X)$ for which

$$\forall x,y \in \tilde{x} \underline{\cup} X: \exists \underset{\cdot}{L} \in L(\tilde{x} \cup X): g(x) - g(y) = \underset{\cdot}{L}(x-y) \tag{3}$$

holds. Let $F: \mathbb{P}X \to \mathbb{R}^n$ be defined by

$$F(Y) := \tilde{x} - Rg(\tilde{x}) + (I - R \cdot L(\tilde{x} \cup Y))(Y - \tilde{x})$$

for all $Y \subset X$, where R is an arbitrary but fixed matrix. If

$$F(X) \subsetneq X \tag{4}$$

holds, then the equation $g(x) = 0$ has one and only one solution $\hat{x}$ in X and $\hat{x} \in F^k(x)$, $k \geq 0$.

Proof [30], [5]

Remark:
Theorem 3 implies that only simple zeros may be enclosed by this method.

The following simple corollary can be directly applied on a computer:

Corollary 1:
Let $\tilde{x}$ be a floating-point vector and Z a floating-point interval vector. $g: Z \underline{\cup} \tilde{x} \to \mathbb{R}^n$ may satisfy a condition (3). Let $H : \mathbb{P}Z \to \mathbb{P}\mathbb{R}^n$,

$$H(Y) := \tilde{x} \diamond R \circledast \textcircled{G} \; (\tilde{x}) \circledast \diamond (I - R \cdot L(\tilde{x} \underline{\cup} \diamond Y)) \circledast (\diamond Y \diamond \tilde{x}) \tag{5}$$

for $Y \in \mathbb{P}Z$, where R is an arbitrary but fixed floating-point matrix. If

$$H(Z) \subsetneqq Z, \tag{6}$$

then the equation $g(x) = 0$ has one and only one solution $\hat{x}$ in Z and $\hat{x} \in H^k (Z)$, $k \geq 0$.

In the case of linear systems $H(Y)$ reduces to

$$H(Y) = \tilde{x} \diamond R \circledast \diamond (Ax - b) \circledast \diamond (I - R * A) \circledast (Y \diamond \tilde{x}) \tag{7}$$

If R is chosen to be an approximation of A^{-1}, it is more likely that condition (6) holds.

Corollary 1 provides the second of the following two basic steps to compute verified bounds for the solution of a system of equations:
1) an approximation step to determine a sufficiently good estimate $\tilde{x}$.
2) an inclusion step to determine the enclosure Z of $\tilde{x}$.

Using corollary 1, a verified inclusion may be obtained by performing the following iteration:

$$Z^{(0)} := \tilde{x}; \; k := 0$$
<u>repeat</u>
   $k := k + 1$
   $Z^{(k)} := H(Z^{(k-1)})$ {H defined by (5) or (7), respectively}
<u>until</u> $Z^{(k)} \subsetneqq Z^{(k-1)}$.

If the termination criterion is satisfied, then by corollary 1 it is verified that the exact solution $\hat{x}$ lies in $Z^{(k)}$ and that no other solution lies between these bounds.

Note that the approximation step which may be performed by usual numerical methods is significant regarding the success of the iteration. Crucial for the accuracy of the result are the computation of ◈ (x̃) in the inclusion step as well as in the approximation. As already mentioned, an evaluation of formula (5) or (7) and test of condition (6) using interval arithmetic would suffice to check all assumptions of our theorem.

Nevertheless, in practice (6) will never happen for intervals with sufficient small diameter if no attempt is made to decrease the diameter of ◈ (x̃) and of the residue ◇(I - RA). For that reason, algorithms which evaluate arbitrary formulae with high accuracy are very helpful. Such evaluators have been developed in the DIAMOND project (work package 2a) [9], [10], [2]. In case of linear systems (7) can be evaluated directly by dot product expressions [35].

The accuracy of the inclusion can be increased if we enclose the error instead of the solution, that means if we represent the solution by a sum of a floating-point vector and an interval vector. If the mantissas of the components of these two vectors do not overlap completely and if we do not carry out the addition in every step we may obtain an inclusion to more than machine precision. This is a way of simulating multiple precision arithmetic. The dot product expressions of PASCAL-SC provide a specific loop construct for these kind of summation.

As an example of an E-method we quote a PASCAL-SC program for the solution of linear systems.

```
procedure solve_lin_system (dim: integer);
   var R, A : rmatrix [1..dim, 1..dim];
        B : imatrix [1..dim, 1..dim];
     xt, b : rvector [1..dim];
xk1, xk, z : ivector [1..dim];
        k : integer;

begin
  mread (input, A); vread (input, b);
  R    := inverse (A);
  xt   := R * b;
  z    := ## (b - A * x);
```

```
B    := ## (mrid(dim) - R * A);
z    := R * z;
xk1 := z;
k    := 0;
repeat
  xk  := xk1;
  xk1 := z + B * xk;
  k   := k + 1
until (xk1 < xk) or (k = 10);
if k < 10 then
begin writeln ('verified inclusion');
      write (xt + xk1);
end
else
      writeln ('inclusion failed')
end
```

The success of the interval iteration is also improved by ϵ-extension, that is an artificial increase of the diameter of the interval.

3. Eigenvalues

a) Complex eigenvalues and eigenvectors for unsymmetric matrices [13], [25].
    The eigenvalue problem

$$Ax = \lambda x$$

with a complex matrix A, a complex vector x, and a complex scalar λ is equivalent to the nonlinear complex system

$$(A - \lambda I)x = 0$$
$$e_k^T x - \xi = 0, \quad \xi \neq 0 \tag{8}$$

where e_k is the k-th unit vector. k is chosen to have the k-th component of the eigenvector $\neq 0$. The approximation is obtained by an

LR or QR algorithm and improved by an iterative refinement procedure using the optimal scalar product.

The verification step applies Brouwer's fixed-point theorem to (8) as described in chapter 2.

b) Eigenvalues of real symmetric matrices [23]

The Jacobi method delivers all eigenvalues of real symmetric matrices simultaneously, irrespective of their multiplicity. A defect correction method is used to improve the orthogonality of the transformation matrices. The inclusion is verified by application of Gerschgorin's theorem.

c) Eigenvalues of Hermitian matrices [8].

Approximations of the eigenvalue are computed by the Jacobi method. These approximations are improved by a Newton method for the system (8). An inclusion is obtained using estimations of the eigenvalues due to Wilkinson [33].

4. The application of theorems on zeros in the complex plane

The following method was originally developed to deal with clusters or multiple zeros of polynomials and is therefore complementary to the methods in [12]. For the description, we summarize some relations from [14].
Let

$$p(z) := a_n z^n + a_{n-1} z^{n-1} + \ldots + a_0 \quad \text{(where } n > 0\text{)}$$

denote a complex polynomial of degree n, even if some of the leading coefficients are zero. We define p*, the reciprocal polynomial of p, by the formula

$$p^*(z) := \overline{a_0} z^n + \overline{a_1} z^{n-1} + \ldots + \overline{a_n}.$$

The reciprocal of a polynomial of degree n is always considered to be of degree n. For $z \neq 0$ the identity

$$p^*(z) = \overline{z^n p(1/\bar{z})}$$

holds.

The polynomial Tp of degree n - 1 is defined by

$$Tp(z) := \overline{a_0} p(z) - a_n p^*(z) = \sum_{k=0}^{n-1} (\overline{a_0} a_k - \overline{a_n} a_{n-k}) z^k$$

is called the Schur transform of p.

We also note that

$$Tp(0) = \overline{a_0} a_0 - a_n \overline{a_n} = |a_0|^2 - |a_n|^2$$

is always a real number.

We define the iterated Schur transforms $T^2 p, T^3 p, \ldots, T^n p$ by

$$T^k p := T(T^{k-1} p), \quad k = 2, 3 \ldots, n,$$

where $T^{k-1} p$ is to be regarded as a polynomial of degree n-k+1 even if its leading coefficient is zero. We set

$$\gamma_k := T^k p(0), \quad k = 1, 2, \ldots, n.$$

Theorem 4:
Let p be a polynomial of degree $n, p \neq 0$. All zeros of p lie outside the closed unit disk $|z| < 1$ if and only if

$$\gamma_k > 0, \quad k = 1, 2, \ldots, n.$$

The calculation of the numbers γ_k is called the Schur-Cohn algorithm. The exact number of zeros in the unit disk is given by the following theorem.

Theorem 5:

Let p be a polynomial of degree n and let the numbers γ_k satisfy $\gamma_k \neq 0$, $k = 1,2,\ldots,n$. If those indices k for which $\gamma_k < 0$ are denoted by k_j, $j = 1,2,\ldots,m$, where $k_1 < k_2 < \ldots < k_m$, then the number h(p) of zeros w of p satisfying $|w| < 1$ (multiple zeros counted with their multiplicity) is given by

$$h(p) = \sum_{j=1}^{m} (-1)^{j-1}(n + 1 - k_j).$$

Together with a bisecting algorithm an interval version of this algorithm can be applied to obtain inclusions for all zeros of a complex polynomial [11]. The number of complex zeros in the unit disk is given by theorem 5. The simple linear transformation $q(z) = p(\rho z + z_0)$ extends this formula to arbitrary disks.

At the beginning of the algorithm all zeros are included in one disk which in turn is included in a diamond. This diamond is successively bisected in order to separate the zeros. If all zeros are separated, a cubically convergent Newton iteration delivers sharp inclusions of the zeros. Otherwise, the bisection is carried on. Note that in the first steps of the bisection it is not necessary to calculate the full Schur-Cohn algorithm. Since a disk may contain a zero if one of the numbers γ_k is less than or equal to zero, the process may be stopped earlier. The complete algorithm uses interval arithmetic. Especially a complex circular interval arithmetic has been implemented.

5. Linear Systems for Sparse Matrices [19]

a) The algorithm for linear systems used for dense matrices in chapter 2 calculates an approximate inverse. Since an inverse of a sparse matrix is not also sparse in general, we must choose other algorithms based on an LU decomposition of A. This can be done in two ways:

   (i) by interval factorization $A = L \cdot U$ and forward/backward substitution:

$$A \cdot x = b \quad \Leftrightarrow \quad \begin{array}{l} L \cdot y = b \\ U \cdot x = b \end{array}$$

(ii) or by approximate factorization $A \approx \check{L} \cdot \check{U}$, the computation of m corrections, $i = 1(1)m$,

$$r^i = \square (b - \sum_{k=0}^{i-1} Ax^k)$$

$$y^i \approx \check{L}^{-1} b$$

$$x^i \approx \check{U}^{-1} y$$

and an interval iteration starting with $X^0 = x^m$

$$\check{L} \diamond \check{U} \diamond X^{i+1} = r^m \diamondsuit [A - \check{L} \cdot \check{U}] \diamondsuit X^i$$

$$x \in \sum_{k=0}^{m} x^k + X^{i+1}$$

$[A - \check{L} \cdot \check{U}]$ is a sharp inclusion of $A - \check{L}\check{U}$ which is easily obtained in each component by one scalar product.

The solution of linear systems with general sparse matrices may also be performed by LU decomposition. Suitable storage methods for the matrices have been investigated. A generic library for sparse matrices has been implemented which exploits the advanced features of the Ada programming language to provide facilities for solving sparse-linear systems [7].

6. <u>Quadrature</u> [16]

In numerical quadrature we may use a standard interpolatory integration formula or one based on Taylor expansion in the general form

$$\int_a^b f(x) \, dx = \underbrace{\sum_{i=1}^n w_i \, f(u_i)}_{S} + \underbrace{c_n \cdot h \, \frac{f^{(p)}(\xi) h^p}{p!}}_{R}$$

where $h = b - a$ and $a < \xi < b$.

We have to enclose the sum $S + R$ where R mainly consists of the Taylor remainder term. u_i, w_i, and c_n are independent of the function f and may therefore be stored in tables.

If the function f is evaluated at the points u_i and stored in a vector, then the computation of S is a simple scalar product. For the computation of R we replace ξ by the interval $[a,b]$ and compute the interval Taylor coefficient.

By a subdivision of the integration interval $[a,b]$ the diameter of R decreases. Then

$$\int_{X_i} f(x)\ dx = \sum_{j=1}^{m} w_{ij}\ f(x_{ij}) + c_{imh_i} \cdot \frac{f^{(p)}(\xi_i)h_i^{\ p}}{p!}$$

holds in each subinterval $X_i \subset [a,b]$. It has been shown [24] that

$$\sum_{i=1}^{k} c_{im}\ \text{diam}\ (X_i)\ F^{(p)}(X_i) \cdot \frac{\text{diam}(X_i)^p}{p!}$$
$$\subseteq c_n\ \text{diam}\ (X_i)\ F^{(p)}(X) \cdot \frac{\text{diam}(X)^p}{p!}.$$

The diameter of the interval Taylor coefficient decreases by a factor of $1/\text{diam}(X)^p$. In addition, the diameter of $F^{(p)}(X)$ overestimates the diameter of $f^{(p)}(X)$ by less than $\text{diam}(X)$ as $\text{diam}(X) \to 0$ if $f^{(p)}(X)$ is continuous. Thus, the gain in accuracy obtained by calculating the error terms over smaller subintervals can be substantial.

A powerful algorithm may be constructed in an adaptive way by subdividing those intervals with the largest diameter. Here we see that interval arithmetic is used to control the flow of a computation, since it can supply us with a measure of accuracy.

We use the Romberg extrapolation method for approximation. Instead of an iterative computation of the T-table, one direct evaluation of a scalar product is computed. Inclusion of an arbitrary high derivative of f is done via automatic differentiation. An adaptive refinement procedure improves both accuracy and performance.

7. Nonlinear Systems [30]

Verification is achieved via application of Brouwer's fixed-point theorem. Since a good approximation is necessary for success, some efforts have been made to improve traditional approximation methods. To illustrate this, we quote Newton's Method as a typical example:

1. Choose $x^{(0)}$;
2. Iterate
   $$x^{(k+1)} := x^{(k)} - f'(x^{(k)})^{-1} \cdot f(x^{(k)}), \quad k = 0,1,2\ldots$$

The iteration in each step is actually performed via

$$
\begin{aligned}
A^{(k)} &:= f'(x^{(k)}) \\
b^{(k)} &:= -f(x^{(k)})
\end{aligned}
$$

Solve $A^{(k)} \cdot d^{(k)} = b^{(k)}$
$$x^{(k+1)} := x^{(k)} + d^{(k)}$$

The points where improvements have been obtained are:
- Evaluation of the function f which is performed with methods developed in work package 2a or by multiple precision arithmetic based on scalar products.
- Solution of the linear system, which exploits the technique sketched in chapters 2 and 5.
- Evaluation of the partial derivatives done by automatic differentiation [24].

Another point for improvement is the starting phase of the iteration. The normal Newton correction $d^{(k)} = f'(x)^{(k)} \cdot f(x)^{(k)}$ could produce a misleading new estimate $x^{(k+1)} := x^{(k)} - d^{(k)}$.
This could be avoided by shortening the correction $d^{(k)}$ by a certain factor μ to satisfy a relation $\| f(x)^{(k)} + \mu d^{(k)} \| < \gamma \| f(x)^{(k)} \|$ [23].

16

In addition, a continuation method helps to find sufficiently close initial guesses.

References

[1] Alefeld, G.; Herzberger, J.: Introduction to Interval Computation. Academic Press, New York, 1983.

[2] Bamberger, L.: Evaluation of Arithmetic Expressions with Standard Functions. DIAMOND Deliverable D2a-2/2, 1988.

[3] Bauch, H. et al.: Intervallmathematik. BSB B.G. Teubner Verlagsgesellschaft, Leipzig 1987.

[4] Böhm, H.: Berechnung von Polynomnullstellen und Auswertung arithmetischer Ausdrücke mit garantierter maximaler Genauigkeit. Dissertation, Universität Karlsruhe 1983.

[5] Böhm, H., Rump, S.M., Schumacher, G.: E-Methods for Nonlinear Problems, in [15], S. 59-80 1987.

[6] Bohlender, G., Rall, L.B., Ullrich, Ch., Wolff von Gudenberg, J.: PASCAL-SC: A Computer Language for Scientific Computation. Academic Press, New York 1987.

[7] Erl, M.J., Klein, W., Wolff von Gudenberg, J.: Verified Results for Linear Systems with Sparce Matrices. DIAMOND Deliverable D3-3, 1988.

[8] Fernando, K.V., Pont, M.W.: Computing Accurate Eigenvalues of a Hermitian Matrix, this volume.

[9] Fischer, H.-C., Haggenmüller, R. und Schumacher, G.: Summary Report on the Tasks. T2a-2, T2a-3. DIAMOND Deliverable D2a-1, 1987.

[10] Fischer, H.-C., Haggenmüller, R.: Formula Evaluation with Standard Functions. DIAMOND Deliverable D2a-1/1, 1987.

[11] Frangen, W.: Verified Inclusion of all Roots of a Complex Polynomial by means of Circular Arithmetic, this volume.

[12] Grüner, K.: Solving Complex Problems for Polynomials and Linear Systems with Verified High Accuracy, in [16], pp. 199-220.

[13] Grüner, K.: Solving the Complex Algebraic Eigenvalue Problem with Verified High Accuracy, this volume.

[14] Henrici, P.: Applied and Computational Complex Analysis, Volume 1. John Wiley & Sons, New York, 1974.

[15] Heuser, H.: Lehrbuch der Analysis, Band II, 3. Auflage. Teubner Verlag, Stuttgart 1986.

[16] Kaucher, E., Kulisch, U., Ullrich, Ch.(Hrsg): Computer Arithmetic - Scientific Computation and Programming Languages. Teubner Verlag, Stuttgart, 1987.

[17] Kaucher, E., Rump, S.M.: E-Methods for Fixed-Point Equations f(x) = x. Computing 28, S. 31-42, 1982.

[18] Kelch, R.: Self-Validating Numerical Quadrature, this volume.

[19] Klein, W.: Verified Results for Linear Systems with Sparse Matrix, this volume.

[20] Kok, J.: The Embedding of Accurate Arithmetic in Ada, in [32].

[21] Kulisch, U., Miranker, W.L.: Computer Arithmetic in Theory and Practice. Academic Press, New York, 1981.

[22] Kulisch, U., Miranker, W.L.: A New Approach to Scientific Computation. Academic Press, New York, 1983.

[23] Lohner, R.: Enclosing all Eigenvalues of Symmetric Matrices, this volume.

[24] Moore, R.: Interval Analysis. Prentice Hall, Englewood Cliffs, N.Y., 1966.

[25] Moynihan, V.: Techniques for Generating Accurate Eigensolutions in ADA, this volume.

[26] Neaga, M., Wolff v. Gudenberg, J.: The PASCAL-SC (Level 2). Compiler Architecture and Implementation Description. DIAMOND Deliverable D1-1, 1988.

[27] Ortega, J.M., Rheinboldt, W.C.: Interative Solution of Nonlinear Equations in Several Variables. Academic Press, Orlando, 1970

[28] Rall, L.B.: Automatic Differentiation: Techniques and Applications. Lecture Notes in Computer Science 120, Springer, Berlin, 1981.

[29] Rump, S.M.: Solving algebraic problems with high accuracy, in [22], pp. 51-120.

[30] Schumacher, G.: Solving Nonlinear Equations with Verification of Results, this volume.

[31] Ullrich, Ch., Wolff von Gudenberg, J.: Different Approaches to Interval Arithmetic, in [32].

[32] Wallis, P.J.L. (ed): Improving Floating-Point Programming, J. Wiley, to appear 1989.

[33] Wilkinson, J.H.: The Algebraic Eigenvalue Problem, Clarendon Press, Oxford 1965

[34] Wolff von Gudenberg, J.: The Embedding of Accurate Arithmetic in PASCAL-SC, in [32].

[35] Wolff von Gudenberg, J.: The Karlsruhe Arithmetic Approach, in [32].

Appendix. The PASCAL-SC Demonstration Package

DIAMOND.PRG

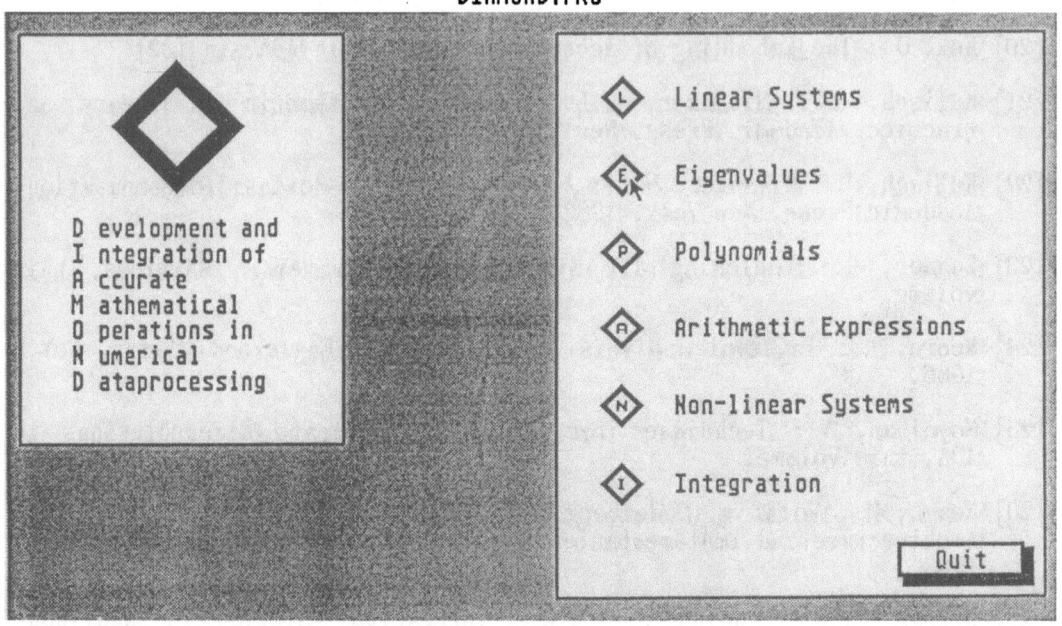

DIAMOND.PRG

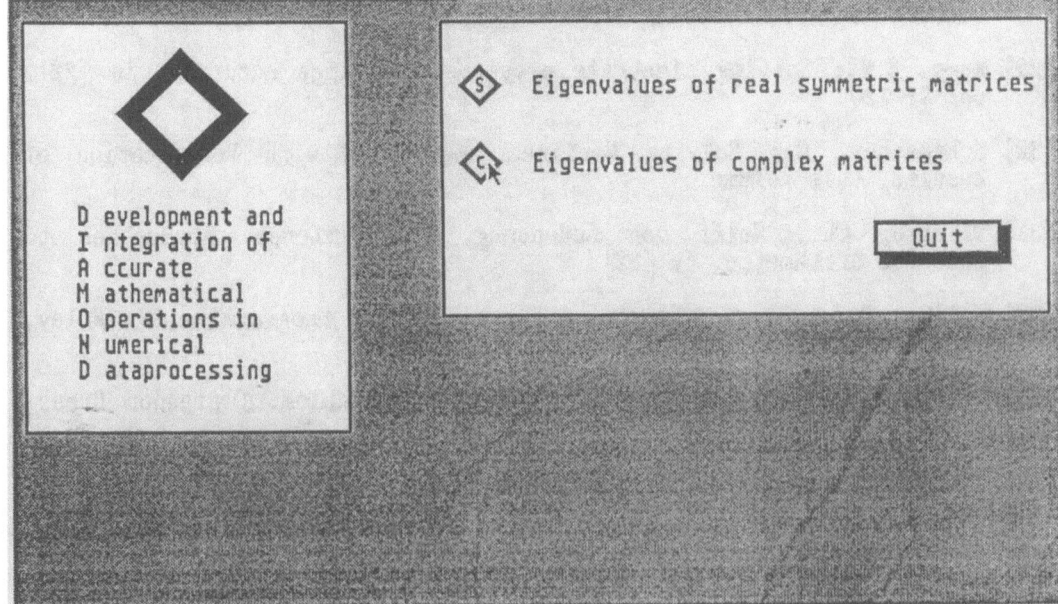

----------- PRECISE COMPLEX EIGENPROBLEM SOLVING -----------

This is a demonstration of precise eigenproblem solving.
For any arbitrary complex matrix eigenvalues and
eigenvectors are computed. The components of the solution
are displayed rounded to the smallest enclosing intervals.
In complex solutions the real part and the imaginary part
are displayed rounded to the smallest enclosing intervals.
The results are algorithmically verified to be correct.


```
E    ENTER       -    Enter input data
B    BROWSE      -    Browse input data
I    INCLUSION   -    Apply precise eigenproblem solving
R    RANDOM      -    Generate random matrix
H    HILBERT     -    Generate hilbert matrix

?    HELP        -    To show options menu
Q    QUIT        -    To terminate precise eigenproblem solving
```

Please enter option. (Enter ? for options menu)
E

------ PRECISE COMPLEX EIGENPROBLEM SOLVING - ENTER ------

First enter the number of rows of the matrix.
Then enter the complex matrix row by row.

Please enter number of rows
3

Please enter the the 3x3 matrix row by row
```
(0,0)      (0,0)        (-13452,13452)
(2378,0)   (0,0)        (0,9512)
(0,0)      (2378,0)     (3363,3363)
```

Generated matrix can be displayed in browse mode

Please enter option. (Enter ? for options menu)
I

------ PRECISE COMPLEX EIGENPROBLEM SOLVING - INCLUSION ------

Enter a number n between 1 and the dimension of the
input matrix.
The approximation and the inclusion of the n-th eigenvalue
and the n-th eigenvector will be displayed.
Enter A to display all.
Additionally the existence of an eigenvalue/eigenvector pair
within the displayed bounds has been algorithmically verified


```
     <n>  1..Dim   - Display n-th eigenvalue/eigenvector pair
     A    ALL      - Display all eigenvalue/eigenvector pairs
     B    BROWSE   - Browse approximations of eigenvalues
     D    DATAFILE - Direct output to additinal outputfile
     S    SCREEN   - Direct output to screen

     ?    HELP     - To show options menu
     Q    QUIT     - To terminate inclusion
```


Output is directed to screen
Please enter option. (Enter ? for options menu)
B

Approximations of eigenvalues

```
     1     ( 6.276950783020E+02,-3.450216281754E+03)
     2     (-3.450216281754E+03, 6.276950783025E+02)
     3     ( 6.185521203449E+03, 6.185521203449E+03)
```


Approximations of eigenvectors

```
     1. eigenvector
     1     ( 1.000000000000E+00, 1.000000000000E+00)
     2     ( 1.613302848011E-01, 7.885248282545E-01)
     3     (-2.564835178234E-01,-4.666184049220E-02)

     2. eigenvector
     1     ( 1.000000000000E+00, 1.000000000000E+00)
     2     ( 1.613302848012E-01,-7.885248282540E-01)
     3     ( 4.666184049220E-02, 2.564835178234E-01)

     3. eigenvector
     1     ( 1.000000000000E+00, 1.000000000000E+00)
     2     ( 1.091552930253E+00,-2.000000000000E-13)
     3     ( 4.598216773311E-01,-4.598216773309E-01)
```

```
Output is directed to screen
Please enter option. (Enter ? for options menu)
A

  1-th eigenvalue:
        1   ( 6.276950783020E+02,-3.450216281754E+03)
( [   6.276950783088E+02,   6.276950783089E+02],
  [ -3.450216281759E+03, -3.450216281758E+03] )

  1-th eigenvector
        1       ( 1.000000000000E+00, 1.000000000000E+00)
( [   1.000000000000E+00,   1.000000000000E+00],
  [   1.000000000000E+00,   1.000000000000E+00] )
        2       ( 1.613302848011E-01, 7.885248282545E-01)
( [   1.613302847991E-01,   1.613302847992E-01],
  [   7.885248282552E-01,   7.885248282553E-01] )
        3       (-2.564835178234E-01,-4.666184049220E-02)
( [ -2.564835178233E-01, -2.564835178232E-01],
  [ -4.666184049278E-02, -4.666184049277E-02] )

  2-th eigenvalue:
        2   (-3.450216281754E+03, 6.276950783025E+02)
( [ -3.450216281759E+03, -3.450216281758E+03],
  [   6.276950783088E+02,   6.276950783089E+02] )

  2-th eigenvector
        1       ( 1.000000000000E+00, 1.000000000000E+00)
( [   1.000000000000E+00,   1.000000000000E+00],
  [   1.000000000000E+00,   1.000000000000E+00] )
        2       ( 1.613302848012E-01,-7.885248282540E-01)
( [   1.613302847991E-01,   1.613302847992E-01],
  [ -7.885248282553E-01, -7.885248282552E-01] )
        3       ( 4.666184049220E-02, 2.564835178234E-01)
( [   4.666184049277E-02,   4.666184049278E-02],
  [   2.564835178232E-01,   2.564835178233E-01] )

  3-th eigenvalue:
        3   ( 6.185521203449E+03, 6.185521203449E+03)
( [   6.185521203449E+03,   6.185521203450E+03],
  [   6.185521203449E+03,   6.185521203450E+03] )

  3-th eigenvector
        1       ( 1.000000000000E+00, 1.000000000000E+00)
( [   1.000000000000E+00,   1.000000000000E+00],
  [   1.000000000000E+00,   1.000000000000E+00] )
        2       ( 1.091552930253E+00,-2.000000000000E-13)
( [   1.091552930253E+00,   1.091552930254E+00],
  [             -7.0E-38,             7.0E-38] )
        3       ( 4.598216773311E-01,-4.598216773309E-01)
( [   4.598216773304E-01,   4.598216773305E-01],
  [ -4.598216773305E-01, -4.598216773304E-01] )


Output is directed to screen
Please enter option. (Enter ? for options menu)
Q
```

----------- PRECISE COMPLEX EIGENPROBLEM SOLVING -----------

This is a demonstration of precise eigenproblem solving.
For any arbitrary complex matrix eigenvalues and
eigenvectors are computed. The components of the solution
are displayed rounded to the smallest enclosing intervals.
In complex solutions the real part and the imaginary part
are displayed rounded to the smallest enclosing intervals.
The results are algorithmically verified to be correct.


```
    E    ENTER      -    Enter input data
    B    BROWSE     -    Browse input data
    I    INCLUSION  -    Apply precise eigenproblem solving
    R    RANDOM     -    Generate random matrix
    H    HILBERT    -    Generate hilbert matrix

    ?    HELP       -    To show options menu
    Q    QUIT       -    To terminate precise eigenproblem solving
```

Please enter option.   (Enter ? for options menu)
H

------ PRECISE COMPLEX EIGENPROBLEM SOLVING - HILBERT ------

A complex hilbert matrix will be generated.
It can be displayed in browse mode.
The matrix is multiplied with a proper constant to
obtain exactly storable input data.
Enter the number of rows of the square matrix.

Please enter number of rows
12

Generated matrix can be displayed in browse mode

Please enter option.   (Enter ? for options menu)
I

------ PRECISE COMPLEX EIGENPROBLEM SOLVING - INCLUSION ------

Enter a number n between 1 and the dimension of the
input matrix.
The approximation and the inclusion of the n-th eigenvalue
and the n-th eigenvector will be displayed.
Enter A to display all.
Additionally the existence of an eigenvalue/eigenvector pair
within the displayed bounds has been algorithmically verified


```
    <n>   1..Dim    - Display n-th eigenvalue/eigenvector pair
     A    ALL       - Display all eigenvalue/eigenvector pairs
     B    BROWSE    - Browse approximations of eigenvalues
     D    DATAFILE  - Direct output to additinal outputfile
     S    SCREEN    - Direct output to screen

     ?    HELP      - To show options menu
     Q    QUIT      - To terminate inclusion
```


Output is directed to screen
Please enter option. (Enter ? for options menu)
10

10-th eigenvalue:
```
      10    ( 1.656161883782E-02, 1.656161883782E-02)
( [  1.665890624666E-02,  1.665890624667E-02],
  [  1.665890624666E-02,  1.665890624667E-02] )
```

10-th eigenvector
```
       1     (-1.539278757519E-05,-1.539278757519E-05)
( [ -1.539266181108E-05, -1.539266181107E-05],
  [ -1.539266181108E-05, -1.539266181107E-05] )
       2     ( 1.155528997415E-03, 1.155528997415E-03)
( [  1.155264527110E-03,  1.155264527111E-03],
  [  1.155264527110E-03,  1.155264527111E-03] )
       3     (-2.098213826196E-02,-2.098213826196E-02)
( [ -2.097354926323E-02, -2.097354926322E-02],
  [ -2.097354926323E-02, -2.097354926322E-02] )
       4     ( 1.564403143528E-01, 1.564403143528E-01)
( [  1.563560757483E-01,  1.563560757484E-01],
  [  1.563560757483E-01,  1.563560757484E-01] )
       5     (-5.672175295372E-01,-5.672175295372E-01)
( [ -5.669122238886E-01, -5.669122238885E-01],
  [ -5.669122238886E-01, -5.669122238885E-01] )
       6     ( 1.000000000000E+00, 1.000000000000E+00)
( [  1.000000000000E+00,  1.000000000000E+00],
  [  1.000000000000E+00,  1.000000000000E+00] )
       7     (-5.795283371802E-01,-5.795283371802E-01)
( [ -5.828206247162E-01, -5.828206247161E-01],
  [ -5.828206247162E-01, -5.828206247161E-01] )
       8     (-5.646343133281E-01,-5.646343133281E-01)
( [ -5.537028491341E-01, -5.537028491340E-01],
  [ -5.537028491341E-01, -5.537028491340E-01] )
```

```
    9       ( 6.543058544602E-01, 6.543058544602E-01)
( [  6.367977120190E-01,  6.367977120191E-01],
  [  6.367977120190E-01,  6.367977120191E-01] )
   10       ( 4.765747570601E-01, 4.765747570601E-01)
( [  4.920834970496E-01,  4.920834970497E-01],
  [  4.920834970496E-01,  4.920834970497E-01] )
   11       (-8.646192158999E-01,-8.646192158999E-01)
( [ -8.719143100888E-01, -8.719143100887E-01],
  [ -8.719143100888E-01, -8.719143100887E-01] )
   12       ( 3.085332881737E-01, 3.085332881737E-01)
( [  3.099592114468E-01,  3.099592114469E-01],
  [  3.099592114468E-01,  3.099592114469E-01] )
```

Output is directed to screen
Please enter option. (Enter ? for options menu)
Q

----------- PRECISE COMPLEX EIGENPROBLEM SOLVING -----------

This is a demonstration of precise eigenproblem solving.
For any arbitrary complex matrix eigenvalues and
eigenvectors are computed. The components of the solution
are displayed rounded to the smallest enclosing intervals.
In complex solutions the real part and the imaginary part
are displayed rounded to the smallest enclosing intervals.
The results are algorithmically verified to be correct.


```
   E    ENTER      -   Enter input data
   B    BROWSE     -   Browse input data
   I    INCLUSION  -   Apply precise eigenproblem solving
   R    RANDOM     -   Generate random matrix
   H    HILBERT    -   Generate hilbert matrix

   ?    HELP       -   To show options menu
   Q    QUIT       -   To terminate precise eigenproblem solving
```


Please enter option. (Enter ? for options menu)
Q

DIAMOND.PRG

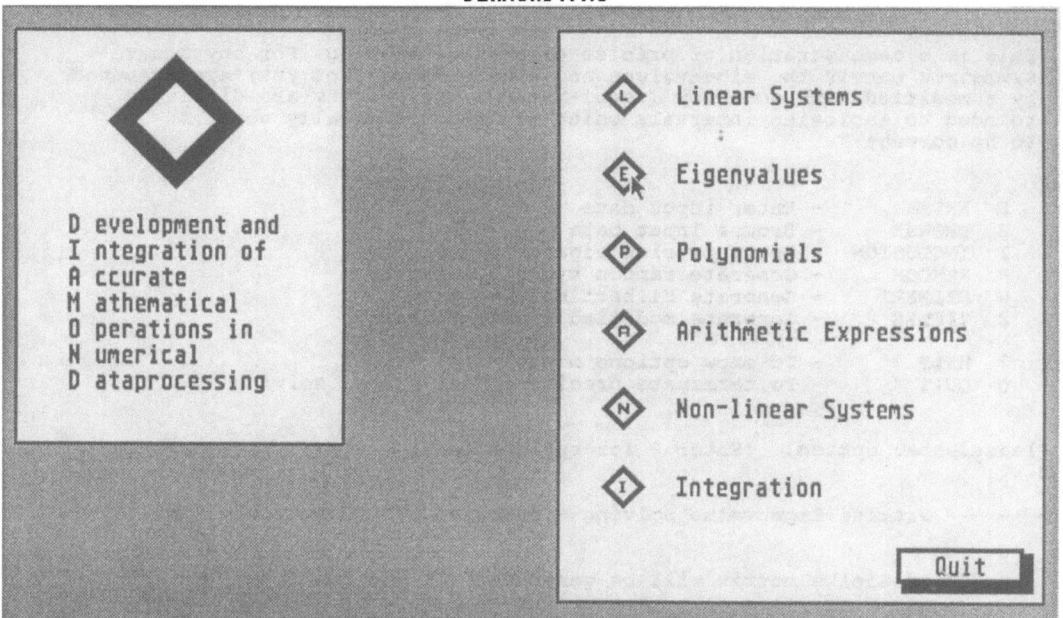

DIAMOND.PRG

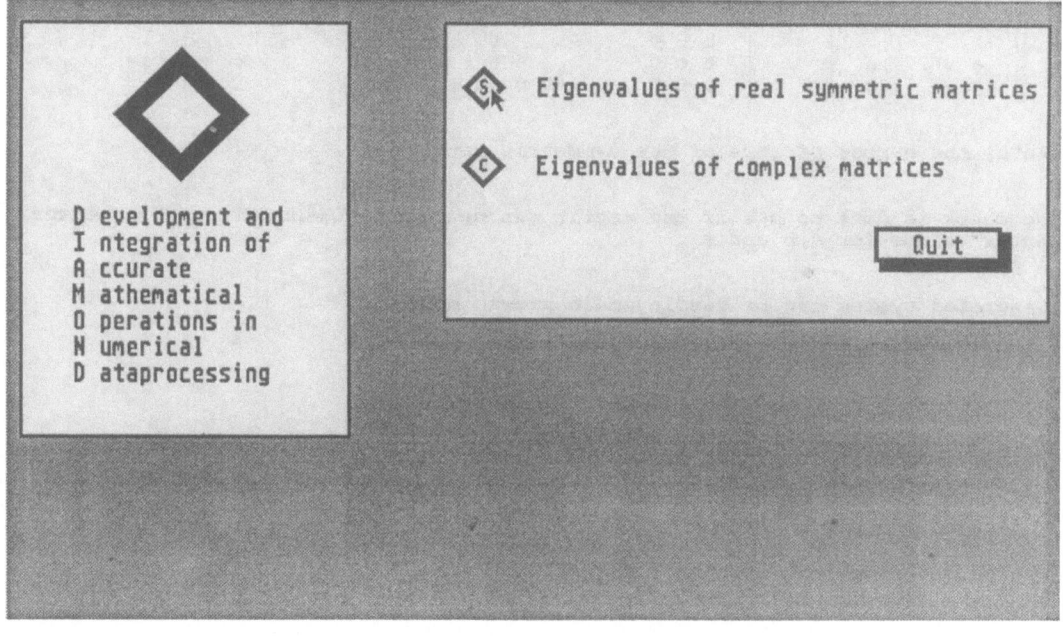

-------- Precise Eigenvalue Solving for Symmetric Matrices -------V1.0-------

This is a demonstration of precise eigenvalue solving. For any square symmetric matrix the eigenvalues and eigenvectors (not yet) are computed by a modified high accuracy Jacobi-Method. The results are displayed rounded to enclosing intervals which are algorithmically verified to be correct.

```
E   ENTER       - Enter input data
B   BROWSE      - Browse input data
I   INCLUSION   - Apply precise eigenvalue solving
R   RANDOM      - Generate random symmetric matrix
H   HILBERT     - Generate Hilbert matrix
Z   ZIELKE      - Generate modified Zielke matrix

?   HELP        - To show options menue
Q   QUIT        - To terminate precise linear system solving
```

Please enter option. (Enter ? for options menu)
Z

-------- Precise Eigenvalue Solving - modified ZIELKE matrix -----V1.0-------

A modified Zielke matrix will be generated. It has the form :

```
| a . . . a |     | b b b . . . b b 0 |     | c               |
| a . . . a |     | b b       b b 0 0 |     |   c             |
| .       . |     | b         . . 0 0 0 |     |     .           |
| .       . |  +  | .     . . . .     . |  +  |       .         |
| .       . |     | . b   . . .       . |     |         .       |
| .       . |     | b b 0 .         0 |     |           .     |
| a . . . a |     | b 0 0         0 0 |     |             c   |
| a . . . a |     | 0 0 0 . . . 0 0 -b |     |               c |
```

Enter the number of rows of the symmetric matrix
4

No check is done to see if the matrix can be computed without rounding errors
Enter values for a,b and c :
1 1E-12 0

Generated system can be displayed in browse mode

Please enter option. (Enter ? for options menu)
B

```
1.000000000001E+00
1.000000000001E+00 1.000000000001E+00
1.000000000001E+00 1.000000000000E+00 1.000000000000E+00
1.000000000000E+00 1.000000000000E+00 1.000000000000E+00 9.999999999990E-01
```

Please enter option. (Enter ? for options menu)
I

```
--------------- Precise Eigenvalue Solving  -  INCLUSION ---------V1.0-------


    B   BROWSE      - Browse inclusions of eigenvalues
    D   DATAFILE    - Direct output to additional outputfile
    S   SCREEN      - Direct output to screen

    ?   HELP        - To show options menu
    Q   QUIT        - To terminate inclusion




   B U S Y  :  A solution is computed
   ---------




--------------- Precise Eigenvalue Solving  -  INCLUSION ---------V1.0-------


    B   BROWSE      - Browse inclusions of eigenvalues
    D   DATAFILE    - Direct output to additional outputfile
    S   SCREEN      - Direct output to screen

    ?   HELP        - To show options menu
    Q   QUIT        - To terminate inclusion

Output is directed to screen
Please enter option.  (Enter ? for options menu)
B


        Inclusion of eigenvalues :

    1    [   4.000000000001E+00,   4.000000000002E+00]
    2    [   5.930703308169E-13,   5.930703308175E-13]
    3    [  -8.430703308173E-13,  -8.430703308172E-13]
    4    [              -5.4E-25,              -4.6E-25]

Additionally the correctness of the displayed results has been
verified by the algorithm.


Output is directed to screen
Please enter option.  (Enter ? for options menu)
Q
```

```
-------- Precise Eigenvalue Solving for Symmetric Matrices -------V1.0-------
```

This is a demonstration of precise eigenvalue solving. For any square
symmetric matrix the eigenvalues and eigenvectors (not yet) are computed
by a modified high accuracy Jacobi-Method. The results are displayed
rounded to enclosing intervals which are algorithmically verified
to be correct.

```
     E   ENTER       - Enter input data
     B   BROWSE      - Browse input data
     I   INCLUSION   - Apply precise eigenvalue solving
     R   RANDOM      - Generate random symmetric matrix
     H   HILBERT     - Generate Hilbert matrix
     Z   ZIELKE      - Generate modified Zielke matrix

     ?   HELP        - To show options menue
     Q   QUIT        - To terminate precise linear system solving
```

Please enter option.   (Enter ? for options menu)

E
```
---------------- Precise Eigenvalue Solving -  ENTER -----------V1.0--------
```

Enter the number of rows of the symmetric matrix A and then the entries
of the lower left triangle of A row by row. The entries must be real numbers

Enter the number of rows of the symmetric matrix
4

Enter the left lower triangle of the matrix row by row
6
4 6
4 1 6
1 4 4 6

Generated system can be displayed in browse mode

Please enter option.   (Enter ? for options menu)
I

```
---------------- Precise Eigenvalue Solving  -  INCLUSION ---------V1.0-------
```

```
     B   BROWSE      - Browse inclusions of eigenvalues
     D   DATAFILE    - Direct output to additional outputfile
     S   SCREEN      - Direct output to screen

     ?   HELP        - To show options menu
     Q   QUIT        - To terminate inclusion
```



```
  B U S Y  :  A solution is computed
  ---------
```

--------------- Precise Eigenvalue Solving - INCLUSION ---------V1.0-------

```
    B   BROWSE      - Browse inclusions of eigenvalues
    D   DATAFILE    - Direct output to additional outputfile
    S   SCREEN      - Direct output to screen

    ?   HELP        - To show options menu
    Q   QUIT        - To terminate inclusion
```

Output is directed to screen
Please enter option. (Enter ? for options menu)
B

      Inclusion of eigenvalues :

```
    1    [   1.499999999999E+01,   1.500000000001E+01]
    2    [  -1.000000000001E+00,  -9.999999999999E-01]
    3    [   4.999999999999E+00,   5.000000000001E+00]
    4    [   4.999999999999E+00,   5.000000000001E+00]
```

Additionally the correctness of the displayed results has been
verified by the algorithm.

Output is directed to screen
Please enter option. (Enter ? for options menu)
Q

-------- Precise Eigenvalue Solving for Symmetric Matrices -------V1.0-------

This is a demonstration of precise eigenvalue solving. For any square
symmetric matrix the eigenvalues and eigenvectors (not yet) are computed
by a modified high accuracy Jacobi-Method. The results are displayed
rounded to enclosing intervals which are algorithmically verified
to be correct.


```
    E   ENTER       - Enter input data
    B   BROWSE      - Browse input data
    I   INCLUSION   - Apply precise eigenvalue solving
    R   RANDOM      - Generate random symmetric matrix
    H   HILBERT     - Generate Hilbert matrix
    Z   ZIELKE      - Generate modified Zielke matrix

    ?   HELP        - To show options menue
    Q   QUIT        - To terminate precise linear system solving
```


Please enter option. (Enter ? for options menu)
Q

DIAMOND.PRG

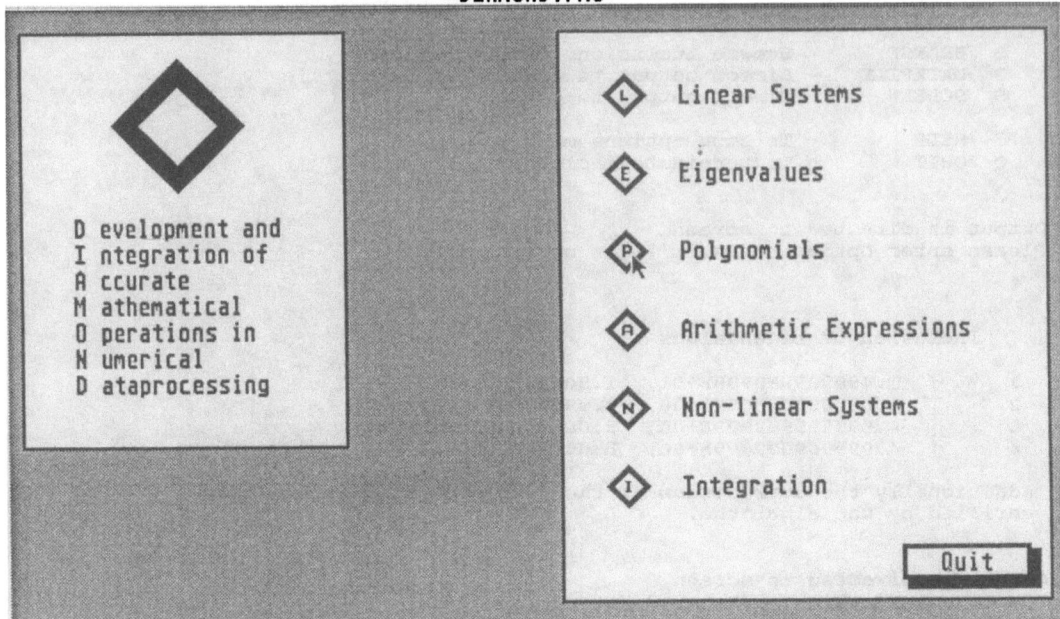

DIAMOND.PRG

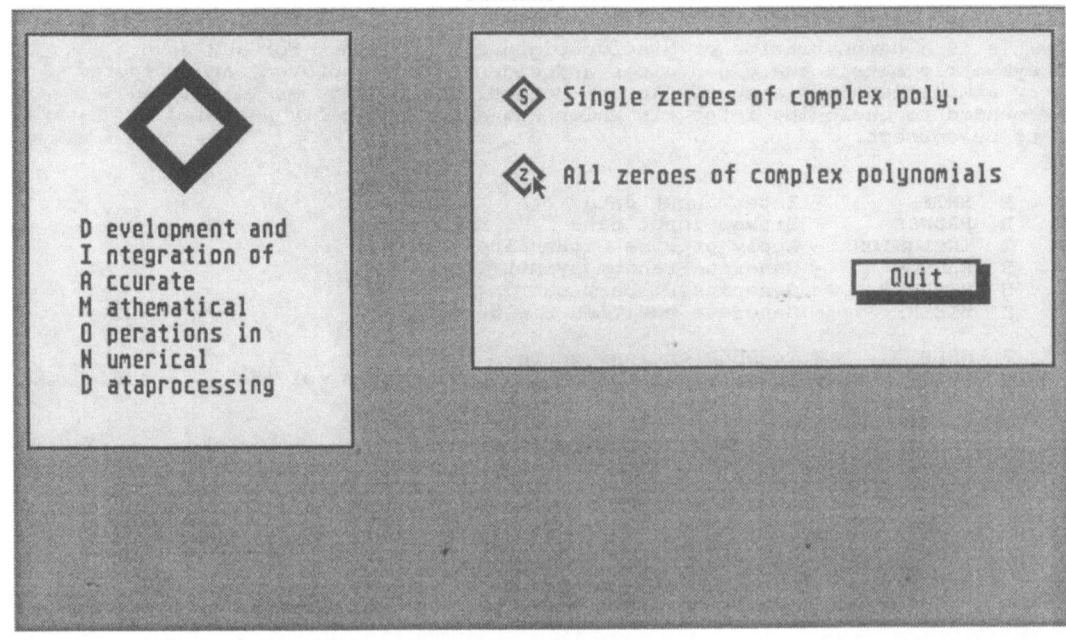

```
----- Verified Inclusion of all roots of a complex polynomial----- V 1.1 ----

    E:    Enter a new polynomial.
    I:    Enter a circle Interval polynomial.
    R:    Generate Random polynomial.
    C:    Construct Polynomial from its roots.
    K:    Show the Keyed in roots.
    B:    Browse last polynomial.
    /:    Scale polynomial roots by divisor 10
    *:    Scale polynomial roots by factor 10.
    T:    Take another bisection grid: Default allow-
          ing 20, T 30, TT 40 bisection steps.
      V:    EValuate polynomial.
      L:    Limit for absolute values of roots.
      G:    Graphical introduction to the method used.
      Z:    Include all roots with maximal precision.
      N:    Apply simultaneous Newton approximation.
      S:    Show the resulting inclusions.
    D:    Direct output to screen / data file.
    ?:    Help: To show options menu.
    Q:    Quit: To terminate.

Please enter option! ("?" for menu.) - Output is directed to data file. R

Please enter polynomial degree!
3

          G e n e r a t i n g   r a n d o m   p o l y n o m i a l .

Polynomial degree : 3.

P(z)=((-2.304914267939E+00,-2.760901177664E-01), 0.000000000000E+00)        +
     ((-9.397935474410E-01, 1.271114867034E+00), 0.000000000000E+00)  z**1 +
     (( 2.100677641640E+00, 2.619320150404E+00), 0.000000000000E+00)  z**2 +
     (( 1.000000000000E+00, 0.000000000000E+00), 0.000000000000E+00)  z**3.


Please enter option! ("?" for menu.) - Output is directed to data file. Z

Wanted accuracy = ? (Write real.) - Default means best obtainable accuracy.
1E-8
```

----- Verified Inclusion of all roots of a complex polynomial----- V 1.1 ----

```
     E:   Enter a new polynomial.
     I:   Enter a circle Interval polynomial.
     R:   Generate Random polynomial.
     C:   Construct Polynomial from its roots.
     K:   Show the Keyed in roots.
     B:   Browse last polynomial.
     /:   Scale polynomial roots by divisor 10
     *:   Scale polynomial roots by factor 10.
     T:   Take another bisection grid: Default allow-
          ing 20, T 30, TT 40 bisection steps.
     V:   EValuate polynomial.
     L:   Limit for absolute values of roots.
     G:   Graphical introduction to the method used.
     Z:   Include all roots with maximal precision.
     N:   Apply simultaneous Newton approximation.
     S:   Show the resulting inclusions.
     D:   Direct output to screen / data file.
     ?:   Help: To show options menu.
     Q:   Quit: To terminate.
```

Please enter option! ("?" for menu.) - Output is directed to data file. S

```
W [ 1]=(( 7.427871846400E-01,-4.406175692800E-01), 5.120000000000E-09) 1
W [ 2]=((-8.488355891200E-01, 1.857346969600E-01), 5.120000000000E-09) 1
W [ 3]=((-1.994629232640E+00,-2.364437273600E+00), 5.120000000000E-09) 1
     Relative error, medium:  4.491968848813E-09;
```

Please enter option! ("?" for menu.) - Output is directed to screen.
G

------- Graphic sketch of the algorithm ---------------------------------

A diamond in the complex plane containig all zeros is computed and then
bisected. The number of zeros in every sub-diamond is estimated.

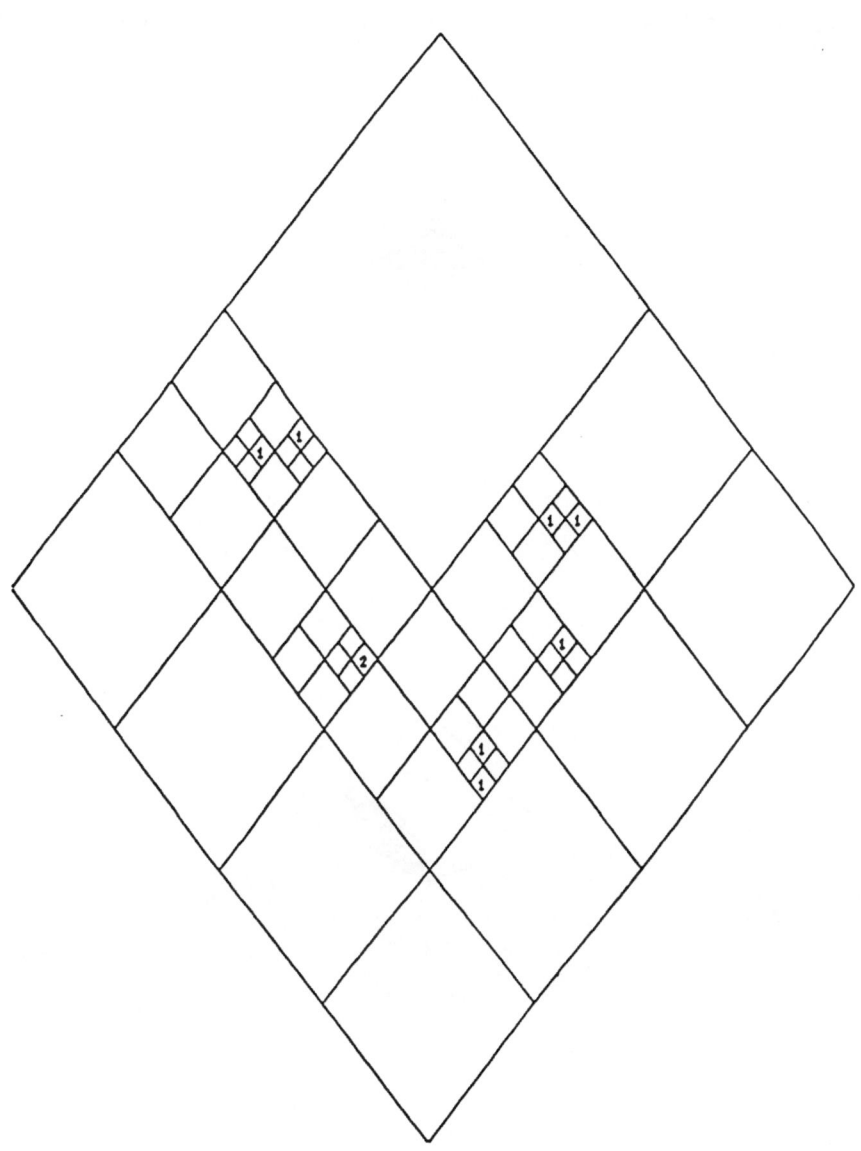

------- Graphic sketch of the algorithm ---------------------------------------

If the zeros are isolated a rapidly converging algorithm determines sharp inclusions. Otherwise bisection continues until one of the following cases occurs.

Actually we compute circles rather than diamonds. The area which contains the zero is filled.

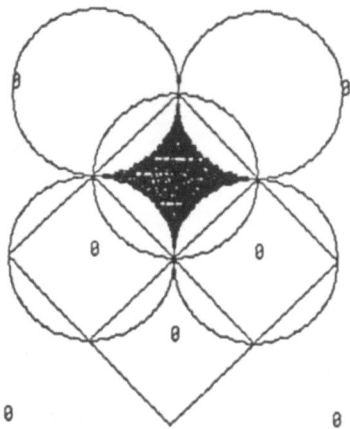

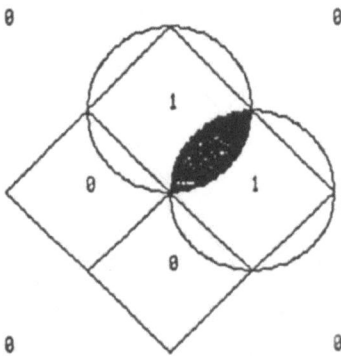

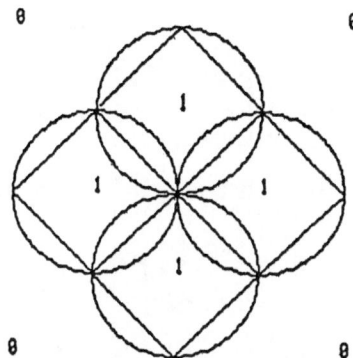

```
----- Verified Inclusion of all roots of a complex polynomial----- V 1.1 ----

    E:    Enter a new polynomial.
    I:    Enter a circle Interval polynomial.
    R:    Generate Random polynomial.
    C:    Construct Polynomial from its roots.
    K:    Show the Keyed in roots.
    B:    Browse last polynomial.
    /:    Scale polynomial roots by divisor 10
    *:    Scale polynomial roots by factor 10.
    T:    Take another model radius: Default value RAD1 allowing 20, RAD2 30,
          RAD3 40 bisection steps; rounding errors may increase with RAD-No.
      V:    EValuate polynomial.
      L:    Limit for absolute values of roots.
      G:    Graph.
      Z:    Include all roots with maximal precision.
      N:    Apply simultaneous Newton approximation.
      F:    Free polynomial roots from scaling.
      S:    Show the resulting inclusions.
    D:  Direct output to screen / data file.
    ?:  Help: To show options menu.
    Q:  Quit: To terminate.

Please enter option! ("?" for menu.) - Output is directed to screen.
Q
```

DIAMOND.PRG

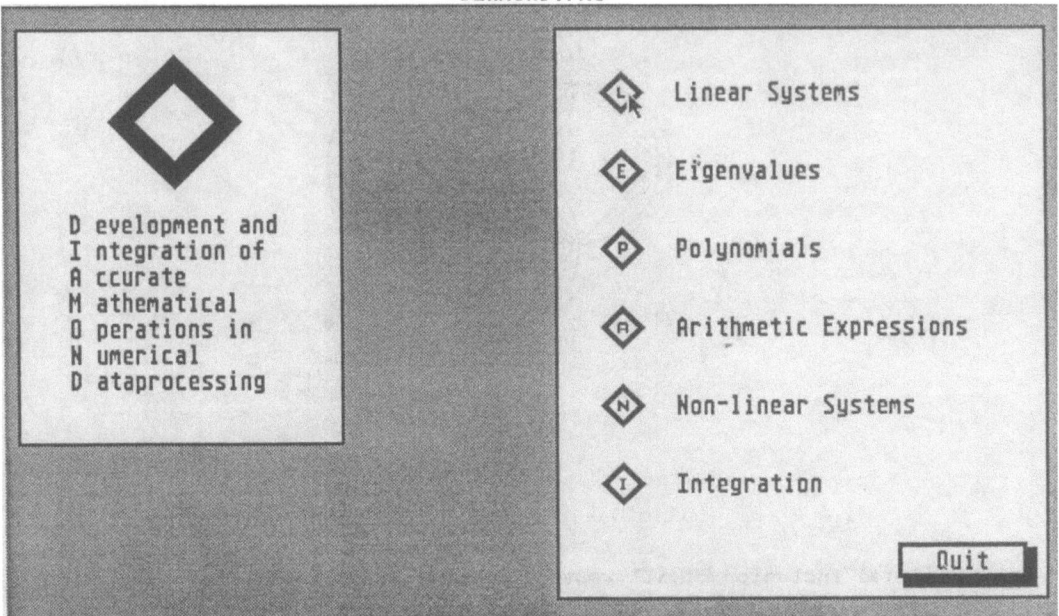

DIAMOND.PRG

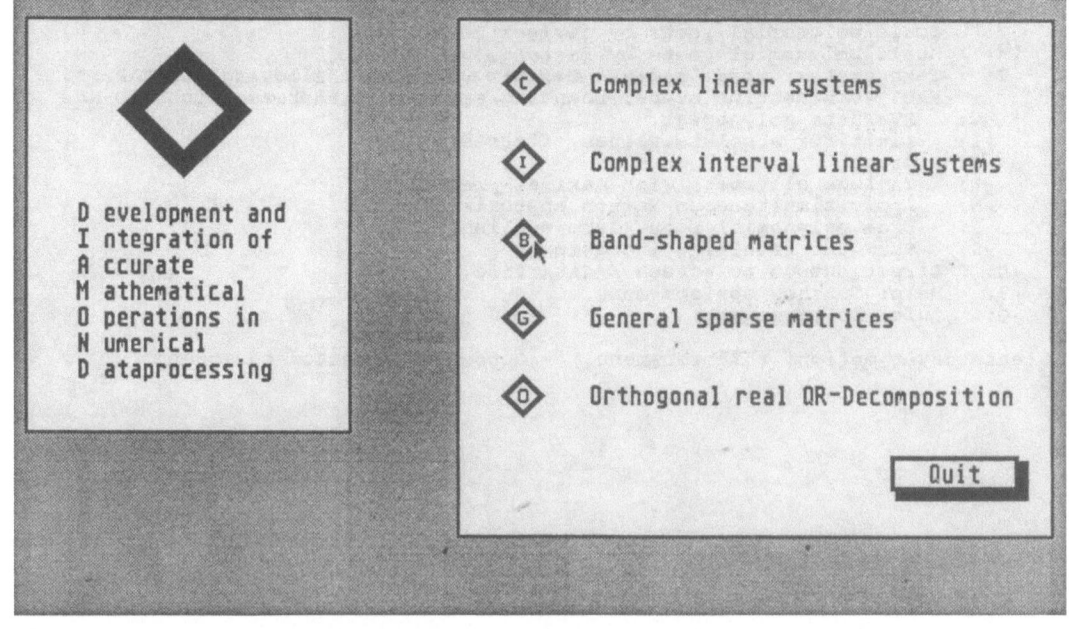

```
----- Band Matrix ------------------------------------------------- V 5.0


   E ENTER          -  Enter input data
   B BROWSE         -  Browse input data
   F FACTORIZATION  -  Factorization of input matrix
   R RIGHT SIDE     -  Enter new right hand side
   I INCLUSION      -  inclusion of linear system
   D DATAFILE       -  Direct output to additional output file
   S SCREEN         -  Output is directed to screen

   ? HELP           -  To show options menu
   Q QUIT           -  To terminate General Sparse Matrix Program


Output is directed to screen
Please enter option.     (Enter ? for options menu)
E




----- Band Matrix ----------------- ENTER ------------------------- V 5.0


Input data for matrix may be read from a file or from the console :

   F FILE           -  Reading from a file
   C CONSOLE        -  Reading from the console


   ? HELP           -  To show options menu
   Q QUIT           -  To terminate ENTER


Please enter option.     (Enter ? for options menu)
C
```

----- Band Matrix ----------------- INPUT ---------------------------- V 5.0

Dimension :
20
Number of upper co - diagonals :
4
Number of lower co - diagonals :
4

Please enter data for A :

" = <value> " assigns <value> to all elements of the (co-) diagonal,
"%" at the beginning of the first upper codiagonal leads to a symmetric
matrix.

Lower co - diagonal 4 :
=1
Lower co - diagonal 3 :
=0
Lower co - diagonal 2 :
=0
Lower co - diagonal 1 :
=1
Main - diagonal :
=-4
Upper co - diagonal 1 :
=1
Upper co - diagonal 2 :
=0
Upper co - diagonal 3 :
=0
Upper co - diagonal 4 :
=1
Change elements ? N

----- Band Matrix -- V 5.0

   E ENTER   - Enter input data
   B BROWSE   - Browse input data
   F FACTORIZATION - Factorization of input matrix
   R RIGHT SIDE   - Enter new right hand side
   I INCLUSION   - inclusion of linear system
   D DATAFILE   - Direct output to additional output file
   S SCREEN   - Output is directed to screen

   ? HELP   - To show options menu
   Q QUIT   - To terminate General Sparse Matrix Program

Output is directed to screen
Please enter option.   (Enter ? for options menu)
R

----- Band Matrix ----------------- RIGHT SIDE ----------------------- V 5.0

 Input data for right hand side may be interval or real :

 I INTERVAL   - Interaval data for right hand side
 R REAL   - Real data for right hand side

Please enter option.
R

----- Band Matrix ----------------- RIGHT SIDE ----------------------- V 5.0

Input data for solution vector may be read from a file or from the console :

 F FILE   - Reading from a file
 C CONSOLE   - Reading from the console

 ? HELP   - To show options menu
 Q QUIT   - To terminate ENTER

Please enter option.   (Enter ? for options menu)
C

----- Band Matrix ----------------- RIGHT SIDE ----------------------- V 5.0

Data for right hand side :

" = <value> " assigns <value> to all elements of the vector
=1

Change elements ? N

```
----- Band Matrix --------------------------------------------------- V 5.0

  E ENTER            -  Enter input data
  B BROWSE           -  Browse input data
  F FACTORIZATION    -  Factorization of input matrix
  R RIGHT SIDE       -  Enter new right hand side
  I INCLUSION        -  inclusion of linear system
  D DATAFILE         -  Direct output to additional output file
  S SCREEN           -  Output is directed to screen

  ? HELP             -  To show options menu
  Q QUIT             -  To terminate General Sparse Matrix Program


Output is directed to screen
Please enter option.     (Enter ? for options menu)
B
```

```
----- Band Matrix ----------------- STRUCTURE ------------------------ V 5.0 ---
```

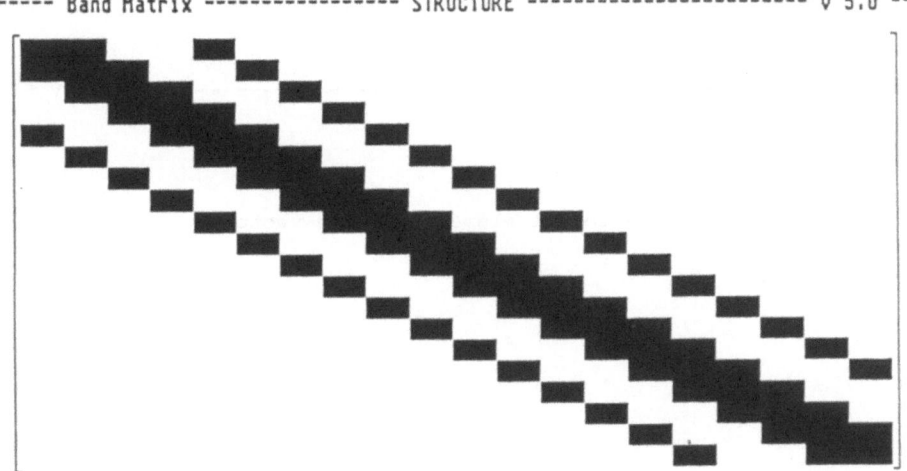

Show elements ? N

```
Output is directed to screen
Please enter option.      (Enter ? for options menu)
F
```

----- Band Matrix ----------------- FACTORIZATION -------------------- V 5.0

 BUSY : Symbolic factorization

 BUSY : Numerical interval factorization

Interval decomposition is possible

----- Band Matrix ----------------- STRUCTURE ------------------------- V 5.0 ---

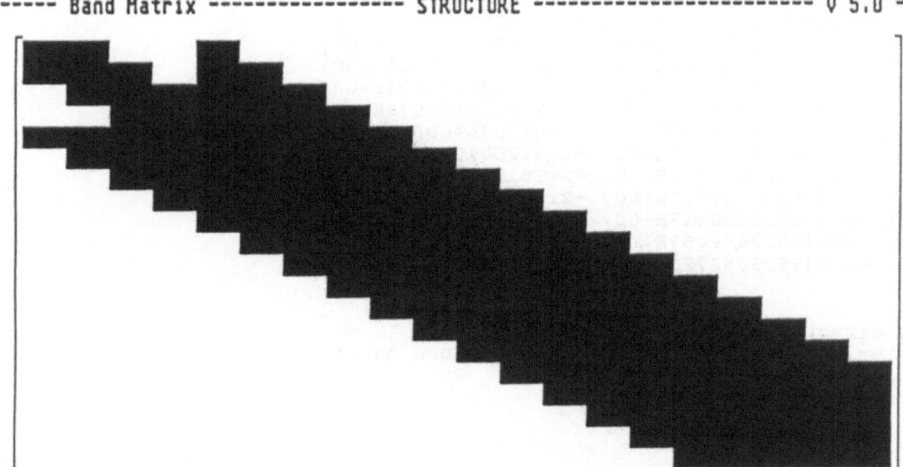

Show elements ? N

Output is directed to screen
Please enter option.   (Enter ? for options menu)
I

----- Band Matrix ----------------- INCLUSION ------------------------- V 5.0

    BUSY : Solving interval system

Vector :
     1   [-1.543623362720E+00, -1.543623362719E+00]
     2   [-2.072932499769E+00, -2.072932499768E+00]
     3   [-2.305515536817E+00, -2.305515536816E+00]
     4   [-2.534305237976E+00, -2.534305237975E+00]
     5   [-3.101560951110E+00, -3.101560951109E+00]
     6   [-3.442591099539E+00, -3.442591099538E+00]
     7   [-3.614824409522E+00, -3.614824409521E+00]
     8   [-3.730144463977E+00, -3.730144463976E+00]
     9   [-3.885724104205E+00, -3.885724104204E+00]
    10   [-3.981046537755E+00, -3.981046537754E+00]
    11   [-3.981046537755E+00, -3.981046537754E+00]
    12   [-3.885724104205E+00, -3.885724104204E+00]
    13   [-3.730144463977E+00, -3.730144463976E+00]
    14   [-3.614824409522E+00, -3.614824409521E+00]
    15   [-3.442591099539E+00, -3.442591099538E+00]
    16   [-3.101560951110E+00, -3.101560951109E+00]
    17   [-2.534305237976E+00, -2.534305237975E+00]
    18   [-2.305515536817E+00, -2.305515536816E+00]
    19   [-2.072932499769E+00, -2.072932499768E+00]
    20   [-1.543623362720E+00, -1.543623362719E+00]

Output is directed to screen
Please enter option.   (Enter ? for options menu)
Q

DIAMOND.PRG

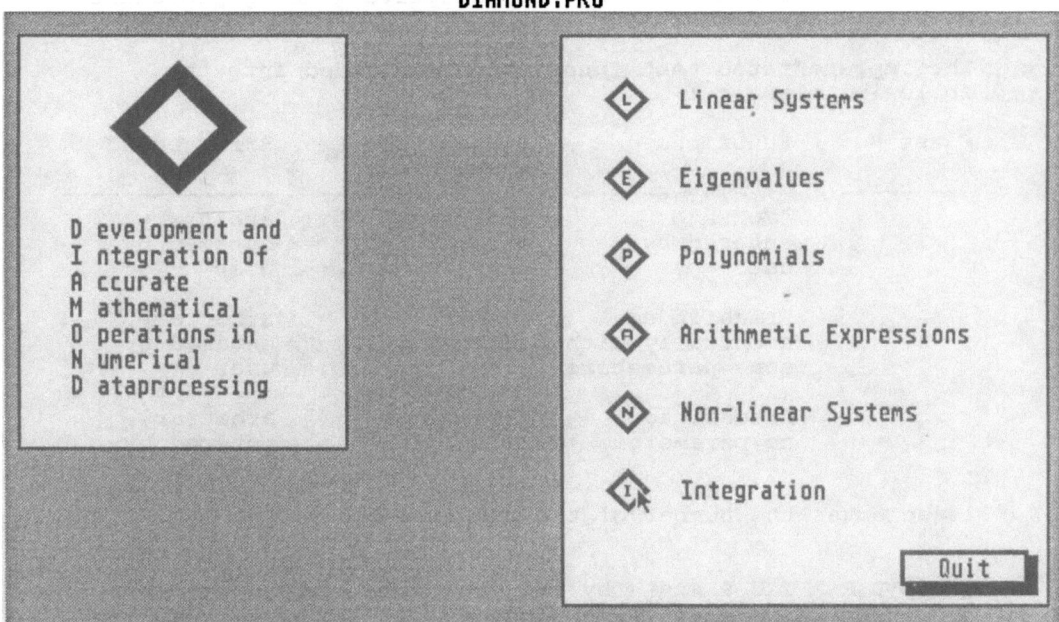

D evelopment and
I ntegration of
A ccurate
M athematical
O perations in
N umerical
D ataprocessing

◇ L Linear Systems

◇ E Eigenvalues

◇ P Polynomials

◇ A Arithmetic Expressions

◇ N Non-linear Systems

◇ I Integration

Quit

Q U A D R A T U R E   by the "adaptive Romberg procedure"

The implementated test functions are divided into the
following classes :

Class No.	Functions	Integration bounds a,b	Error bound eps
1	"JOKER" entered by user	arbitrary, entered by user	arbitrary, entered by user
2	Predefinded, eventually with some parameters	arbitrary, entered by user	arbitrary, entered by user
3	Predefined, no parameters	Predefined	arbitrary, entered by user

Please enter the number of the function class :
2

Q U A D R A T U R E   by the "adaptive Romberg procedure"

You may choose one of the following types of functions :

```
type  1:   f(x) := (a1*x + a2) ^ n     //with n<>-1)
type  2:   f(x) := 1 / (a1*x + a2)
type  3:   f(x) := EXP(x)*(1 + 1/x + LN(x))
type  4:   f(x) := a1*SIN (x) + a2*COS(x)
type  5:   f(x) := TAN (x)
type  6:   f(x) := COT (x)
type  7:   f(x) := 1 / SQR(COS( x ))
type  8:   f(x) := 1 / SQR(SIN( x ))
```

Please enter the number of the function and press RETURN.

>>> for more function types enter " 0 " !
4

Please enter values for the real constants a1 and a2.
1 0

Please enter the integration interval.

 lower bound =
-0.1
 upper bound =
0.1

Please enter a value for the maximum error bound.
1E-11

Do you want comments ? - Y/N
N

```
BUSY : Integrating...

        R U N T I M E - I N F O R M A T I O N
        =======================================
```

Step	msecs	# function evaluations
Segmentation and computation of derivatives	2355	
1. subinterval	505	5
2. subinterval	505	5
all subintervals	1035	10
SUMMATION	5	none

Please press RETURN to continue

```
                        R E S U L T
-----------------------------------------------------------------
```

The result of the quadrature of the function

```
type  4:   f(x) := a1*SIN (x) + a2*COS(x)
    with   a1 =  1.000000000000E+00
           a2 =  0.000000000000E+00
```

from -1.000000000000E-01   up to 1.000000000000E-01

is :

```
integral value        = [            -1.4E-14,         1.4E-14]
value for comparison = [            -1.0E-13,         1.0E-13]
(evaluation of primitive function)
```

The computed result is a better inclusion than the evalution
of the primitive function

```
received absolute error =  2.631677999412E-14
required relative error =  1.000000000000E-09
```

Repeat computation with another error bound ? Y/N
N

Do you want to enter a new integration interval for
the same function ? Y/N
N

Do you want to choose a new function ? Y/N
Y

Q U A D R A T U R E   by the "adaptive Romberg procedure"

The implementated test functions are divided into the
following classes :

Class No.	Functions	Integration bounds a,b	Error bound eps
1	"JOKER" entered by user	arbitrary, entered by user	arbitrary, entered by user
2	Predefinded, eventually with some parameters	arbitrary, entered by user	arbitrary, entered by user
3	Predefined, no parameters	Predefined	arbitrary, entered by user

Please enter the number of the function class :
1

>>>>> The implemented function is
      f(x) := SQR(SIN(X))+SQR(COS(X))

Please enter the integration interval.

 lower bound =
0
 upper bound =
1

Please enter a value for the maximum error bound.
1E-11

Do you want comments ? - Y/N
N

BUSY : Integrating...

        R U N T I M E - I N F O R M A T I O N
        ======================================

Step	msecs	# function evaluations
Segmentation and computation of derivatives	4270	
1. subinterval	1920	17
2. subinterval	2160	17
all subintervals	4095	34
SUMMATION	5	none

Please press RETURN to continue

R E S U L T

The result of the quadrature of the function

"JOKER" :   f(x) := SQR(SIN(X))+SQR(COS(X))

from 0.000000000000E+00   up to 1.000000000000E+00

is :

integral value   = [9.999999999989E-01, 1.000000000002E+00]
value for comparison = [1.000000000000E+00, 1.000000000000E+00]
(evaluation of primitive function)

The computed result is an inclusion of the evaluation
of the primitive function

received relative error = 3.100000000004E-12
required relative error = 1.000000000000E-11

Repeat computation with another error bound ? Y/N
N

Do you want to enter a new integration interval
for the same function ? Y/N
N

Do you want to choose a new function ? Y/N
Y

Q U A D R A T U R E   by the "adaptive Romberg procedure"

The implementated test functions are divided into the
following classes :

Class No.	Functions	Integration bounds a,b	Error bound eps
1	"JOKER" entered by user	arbitrary, entered by user	arbitrary, entered by user
2	Predefinded, eventually with some parameters	arbitrary, entered by user	arbitrary, entered by user
3	Predefined, no parameters	Predefined	arbitrary, entered by user

Please enter the number of the function class :
2

Q U A D R A T U R E   by the "adaptive Romberg procedure"

You may choose one of the following types of functions :

```
type  1:   f(x) := (a1*x + a2) ^ n     //with n<>-1)
type  2:   f(x) := 1 / (a1*x + a2)
type  3:   f(x) := EXP(x)*(1 + 1/x + LN(x))
type  4:   f(x) := a1*SIN (x) + a2*COS(x)
type  5:   f(x) := TAN (x)
type  6:   f(x) := COT (x)
type  7:   f(x) := 1 / SQR(COS( x ))
type  8:   f(x) := 1 / SQR(SIN( x ))
```

Please enter the number of the function and press RETURN.

>>> for more function types enter " 0 " !
2

Please enter values for the real constants a1 and a2.
1 0

Please enter the integration interval.

 lower bound =
0.5
 upper bound =
1.5

Please enter a value for the maximum error bound.
1E-11

Do you want comments ? - Y/N
N

BUSY : Integrating...

        R U N T I M E - I N F O R M A T I O N
        =======================================

Step	msecs	# function evaluations
Segmentation and computation of derivatives	420	
1. subinterval	1350	65
2. subinterval	655	33
all subintervals	2015	98
SUMMATION	10	none

Please press RETURN to continue

R E S U L T

The result of the quadrature of the function

type 2:   f(x) := 1 / (a1*x + a2)
    with   a1 = 1.000000000000E+00
           a2 = 0.000000000000E+00

from 5.000000000000E-01   up to 1.500000000000E+00

is:

integral value   = [1.098612288666E+00,   1.098612288670E+00]
value for comparison = [1.098612288668E+00,   1.098612288669E+00]
(evaluation of primitive function)

The computed result is an inclusion of the evaluation
of the primitive function !

received relative error = 3.640956906515E-12
required relative error = 1.000000000000E-11

Repeat computation with another error bound ? Y/N
N

Do you want to enter a new integration interval
for the same function ? Y/N
N

Do you want to choose a new function ? Y/N
N

DIAMOND.PRG

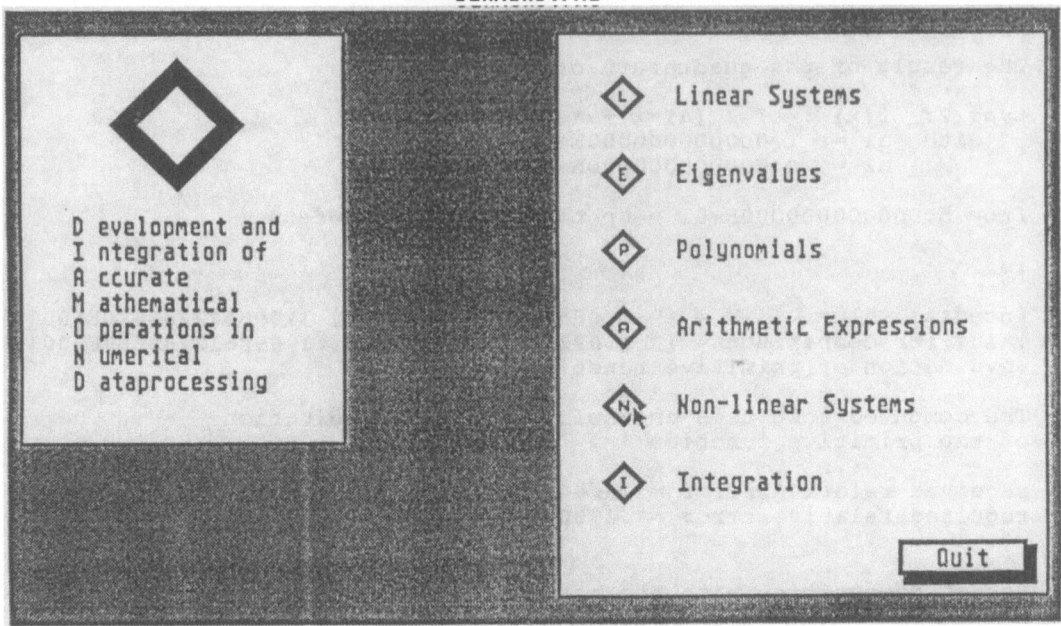

Contents of the input file :

```
@functions
A[1,1]*X[1] + A[1,2]*X[2] + A[1,3]*X[3]*X[3] - B[1];
A[2,1]*X[1] + A[2,2]*X[2]*X[2] + A[2,3]*X[1]*X[3] - B[2];
A[3,1]*X[1]*X[2] + A[3,2]*X[3] - B[3];
@unknowns
X[1] := 0
X[2] := 0
X[3] := 0
@parameters
A[1,1] := 3
A[1,2] := 1
A[1,3] := 2
A[2,1] := -3
A[2,2] := 5
A[2,3] := 2
A[3,1] := 25
A[3,2] := 20
B[1] := 3
B[2] := 1
B[3] := -12
@end
```

```
------------------ Solver for systems of nonlinear equations ---------- V1.1


    Please, select:


        C       Solve system on input file

        F       Start continuation method

        O       Set options

        H       Display help information

        Q       Quit program


Command => H
```

```
---------------------------- Help ----------------------- Page 1 of 5


    This program tries to solve a system of nonlinear algebraic equations
    of the form

                    f[1](x[1],...,x[n]) = 0
                            .
                            .
                            .
                    f[n](x[1],...,x[n]) = 0

    where n is the degree of the system and the functions f[i] are formulas
    consisting of

        - the unknowns x[1], x[2], ..., x[n],
        - the operations +,-,*,/ and parenthesis ( and ),
        - explicit constants such as 0.314e+1,
        - named constants (parameters).


                                        *** Press RETURN
```

1. How to put your system in

    The user has to provide an input file named 'INPUT.NLS' in the same
    directory as 'NLSS.TOS'. A sample input file would look as follows:

    ```
    ------ Top of File ------
    @functions
    x[1]*x[1] + x[2]*x[2] - r;        <-- This is the definition of f[1]
    x[1]*x[1] - x[2];                 <-- This is the definition of f[2]
    @unknowns
    x[1] = 1.0                        <-- initial guess for x[1] is 1.0
    x[2] = 1.0
    @parameters
    r = 2.3                           <-- parameter r has value 2.3
    @end
    ------ End of File ------
    ```

    Each function definition has to start in a new line but may take more
    than one line; the end is marked by ';'. Parameter definition is
    optional. In names, '[', ']', and commas are allowed (e.g. A[2,4]);
    the total length of names is 12.
                                                         *** Press RETURN

2. Normal execution

    - Prepare an input file INPUT.NLS as explained above

    - Start NLSS.TOS and select item 'C'

    - The approximation step followed by a verification step is started.
      A protocol of the whole session can be found in the file OUTPUT.NLS

3. What to do if inclusion fails

    You can choose one of the following three improvements to recover a
    possible failing of the algorithm:

    - Restart with higher precision (first item in 'Options' panel)

    - Start continuation method (select item 'F')

    - Change one or more of the program parameters displayed in the
      'Options' panel
                                                         *** Press RETURN

4. Explanation of the program parameters

- Bound for relative error (EPS):
  The approximation stops if for two succeeding iterates x[i-1] and
  x[i] and for all j in {1,...,n} it holds that

  $$| x_j[i-1] - x_j[i] | <= EPS * | x_j[i] |$$

- Maximum iteration count (KMAX):
  To prevent infinite looping the iteration stops with no success
  if the number of iterations reaches the maximum iteration count KMAX.

- Maximum number of starting point variations (QMAX):
  If the iteration stops with no success some variations of the
  initial guesses will be tried. QMAX is the maximum number of trials.

- Factor for starting point variation (FACTOR):
  The q-th variation has the form

  $$x_i[new] := x_i[old] * sin(q*i) * FACTOR, 1<=i<=n$$
  *** Press RETURN

- Maximum number of Gauss-Seidel steps (GMAX):
  Sometimes systems can be be solved by an iterated iteration where
  in the i-th step of the inner iteration the i-th equation is solved
  for the i-th unknown and all other unknowns are to be considered as
  constants. This process needs some assumptions about the structure
  of the system (ref. literature) and can be applied only to special
  systems. GMAX is the maximum number of outer iterations.

- Factor for norm decreasing (MUE):
  To prevent that the iteration runs amok due to singularities in the
  near of the starting values, two succeeding iterates x[i-1] and x[i]
  must satisfy

  (*) || f(x[i]) || < MUE * || f(x[i-1]) || .

  If this doesn't hold a bisection of the correction x[i]-x[i-1] is
  started until (*) is satisfied.

- Maximum number of bisection steps (LMAX):
  If (*) could not be achieved at least LMAX bisections will be
  performed.
  *** Press RETURN

```
----------------- Solver for systems of nonlinear equations ---------- V1.1


     Please, select:


          C     Solve system on input file

          F     Start continuation method

          O     Set options

          H     Display help information

          Q     Quit program


  Command => C



Approximation after 15 step(s):

X[1]        =   1.050753840624E-03
X[2]        =   8.297803677357E-02
X[3]        =  -1.744790125086E-02

Iteration not convergent!

Press RETURN...



----------------- Solver for systems of nonlinear equations ---------- V1.1


     Please, select:


          C     Solve system on input file

          F     Start continuation method

          O     Set options

          H     Display help information

          Q     Quit program


  Command => F
```

----------------------------- Continuation Method --------------------- V1.1

                zero of g(x,t) := f(x) + (t-1)*f(x0), t in [0,1]

                D   Show results of intermediate steps   (OFF)

                Z   Number of intersections of [0,1]   (10)

                L   Lower bound for step size   (1.0E-01)

                S   Start continuation method

Command => S

t = 1.000000000000E-01   Ok
t = 2.000000000000E-01   Ok
t = 3.000000000000E-01   Ok
t = 4.000000000000E-01   Ok
t = 5.000000000000E-01   Ok
t = 6.000000000000E-01   Ok
t = 7.000000000000E-01   Ok
t = 8.000000000000E-01   Ok
t = 9.000000000000E-01   Ok
t = 1.000000000000E+00   Ok

Approximation after 4 step(s):

X[1]   = 2.900523457550E-01
X[2]   = 6.874306252633E-01
X[3]   = -8.492385817519E-01

Inclusion after 1 step(s):

X[1]   = [2.90052345754E-01,   2.90052345756E-01]
X[2]   = [6.874306252630E-01, 6.874306252638E-01]
X[3]   = [-8.49238581753E-01, -8.49238581751E-01]

Estimation for the condition: 1.104E+01

Press RETURN...

```
------------------- Solver for systems of nonlinear equations ----------- V1.1


     Please, select:


          C      Solve system on input file

          F      Start continuation method

          O      Set options

          H      Display help information

          Q      Quit program


 Command => O
```

```
------------------------------- Options ------------------------- V1.1
          L      Precision (number of mantissas)       (1)
          S      Old initial guesses at restart?       (NO)

          *** Approximation            ***

          R      Bound for relative error              ( 1.0E-10)
          M      Maximum iteration count               (15)

          *** Starting phase of iteration ***

          G      Maximum number Gauss-Seidel steps     (0)
          V      Max. number starting point variations (2)
          P      Factor for starting point variation   ( 1.1E+00)
          H      Maximum number bisection steps        (10)
          N      Factor for norm decreasing            ( 5.0E-01)


          Q      Quit options


 Command => L
 Number of mantissas: 2
```

```
-------------------------------- Options ---------------------------- V1.1

      L      Precision (number of mantissas)        (2)
      S      Old initial guesses at restart?        (NO)

      *** Approximation              ***

      R      Bound for relative error               ( 1.0E-10)
      M      Maximum iteration count                (15)

      *** Starting phase of iteration ***

      G      Maximum number Gauss-Seidel steps      (0)
      V      Max. number starting point variations  (2)
      P      Factor for starting point variation    ( 1.1E+00)
      H      Maximum number bisection steps         (10)
      N      Factor for norm decreasing             ( 5.0E-01)


      Q      Quit options


Command => Q


------------------ Solver for systems of nonlinear equations ---------- V1.1


    Please, select:


      C      Solve system on input file

      F      Start continuation method

      O      Set options

      H      Display help information

      Q      Quit program


Command => C
```

Approximation after 1 step(s):

```
X[1]        =   2.900523457550E-01
X[2]        =   6.874306252634E-01
X[3]        =  -8.492385817518E-01
```

Inclusion after 1 step(s):

```
X[1]        = [   2.900523457549E-01,   2.900523457551E-01]
X[2]        = [   6.874306252633E-01,   6.874306252635E-01]
X[3]        = [  -8.492385817519E-01,  -8.492385817517E-01]
```

Estimation for the condition: 1.104E+01

Press RETURN...

------------------ Solver for systems of nonlinear equations ---------- V1.1

        Please, select:

            C   Solve system on input file

            F   Start continuation method

            O   Set options

            H   Display help information

            Q   Quit program

    Command =>   Q

Solving the Complex Algebraic Eigenvalue Problem with Verified High Accuracy

Diamond Deliverable D3-1, part 2

K. Grüner
Universität Karlsruhe

Abstract:

An efficient method is presented to include separated complex eigenvalues and eigenvectors by narrow bounds. First, some mathematical foundations and an algorithm for a good floating-point approximation are given. Then, the verification algorithm is described which has the property that the true solution is enclosed by computed bounds with high accuracy. The existence and uniqueness of true solution within these bounds are simultaneously proven; that is, the stated method yields results with verified high accuracy.

1. Introduction

The aim of this paper is to state an efficient algorithm, approved in practice, for the algebraic eigenvalue problem. The result of the algorithm encloses the true solution by computed bounds with high accuracy. If an enclosure can be computed, the existence and uniqueness of the true solution are proved.

The basis for the implementation of the algorithm is the availability of an arithmetic with maximum accuracy for all real numerical spaces, as defined in [4]. Such an optimal arithmetic is not easily implemented; refer to the contributions [2], [3], [4] and [5] concerning this fact.

Let R denote the real floating-point system, VR the set of vectors and MR the set of matrices over R. IT represents the set of intervals with bounds in T where $T \in \{R, VR, MR\}$. The complex floating-point system is denoted by C. $\square$ is a monotone antisymmetric rounding and ∇ and $\triangle$

are the downwardly and upwardly directed roundings, respectively. Let
$* \in \{+, -, \cdot, /\}$, then be the real floating-point operation in the machine
representable spaces are denoted by ⊞, ▽, and ▲ for these roundings. The
rounding to the smallest enclosing interval is marked by ◇, and ◈ with
$* \in \{+, -, \cdot, /\}$ represents the interval operations in the corresponding
space.

Since all operations are defined with maximum accuracy (cf [4]) for all
numerical spaces, the following error estimation is valid as long as no
under or overflow occurs:

$$| x * y - x \circledast y | \leq \epsilon \cdot | x \circledast y |$$

with $\bigcirc \in \{\square, \nabla, \triangle, \diamond\}$ and $\epsilon = b^{1-1}$, where b denotes the base of the
floating-point system and 1 the mantissa length. This accuracy estimation
is also valid for the spaces of vectors and matrices. Therefore, it is
particularly important that the real scalar product $\sum_{i=1}^{n} x_i y_i$ with
$x_i, y_i \in R$ is realized with maximum accuracy. An arithmetic satisfying
these requirements is available in PASCAL-SC or via the ACRITH and ARITHMOS
subroutine library.

Now we explain the term "maximum accuracy": A real-floating point number
represents a result of maximum accuracy if no further number of the same
real floating-point system lies between the computed and the exact result.
A real result interval of maximum accuracy is an interval which contains
the exact result and statisfies the following requirements: It is a point
interval or an interval whose bounds are neighboured by real floating-point
numbers or an interval containing in its interior at most one real
floating-point number. Corresponding statements also hold for the complex
case.

2. Mathematical Foundations

The algebraic eigenvalue problem

$$A \cdot x = \lambda x$$

with $A \in M_n\mathbb{C}$, $x \in V_n\mathbb{C}$ and $\lambda \in \mathbb{C}$ is equivalent to the nonlinear complex system

$$(A - \lambda \cdot E) \cdot x = 0$$
$$e_k^T \cdot x - \xi = 0, \quad 0 \neq \xi \in \mathbb{C}$$

where E denotes the identity matrix and $e_k \in V_n\mathbb{C}$ the kth identity vector.

We first aim, therefore, at obtaining an algorithm for the inclusion of nonlinear complex systems. For the proof of this central theorem some other theorems are necessary. They are presented here without being proven.

Theorem 1: (Fixed-point theorem of Schauder)
   Let T be a normed linear space of the dimension n, $K \subseteq T$ convex and $G \subseteq K$ compact and nonempty. Then every continuous map $A : K \longrightarrow G$ has at least one fixed-point.

In the special case that $T = V_n \mathbb{C}$, it follows immediately that

Theorem 2:
   Let $f : V_n\mathbb{C} \longrightarrow V_n\mathbb{C}$ be a continuous map, $X \subseteq V_n\mathbb{C}$ a nonempty, convex and compact set and

$$f(x) \subseteq X$$

   is valid. Then the equation $f(z) = z$ has at least one solution $z \in X$.

If the matrix $A \in M_n\mathbb{C}$ has the eigenvalues λ_i, $i = 1(1)n$ then the spectral radius $\rho(A)$ is defined by

$$\rho(A) := \max_{1 \leq i \leq n} |\lambda_i|.$$

A consequence of Theorem 2 for the complex interval computation is the

Theorem 3:

Let $A \in M_n\mathbb{C}$, $z \in V_n\mathbb{C}$ and $X \in IV_n\mathbb{C}$. If

$$z + A \cdot X \subseteq X,$$

then $\rho(A) < 1$ is valid.

An immediate conclusion is the

Theorem 4:

Let $A \in M_n\mathbb{C}$, $z \in V_n\mathbb{C}$ and $X \in IV_n\mathbb{C}$. If

$$z + (E - A) \cdot X \subseteq X,$$

then the complex matrix A is nonsingular.

An important auxiliary for proof in the real case is the mean-value theorem. It is not valid in the complex case, but it is possible to make do with the following theorem:

Theorem 5:

Let $G \in PV_n\mathbb{C}$ be a convex and nonempty set, $f: G \longrightarrow \mathbb{C}$ a holomorphic function and z, $z_0 \in G$. Then there exist two complex vectors ξ_1, ξ_2 in the set

$$S := \{z_0 + t \ (z - z_0) \ | \ 0 \leq t \leq 1, \ t \in \mathbb{R}\}$$

with the property

$$f(z) = f(z_0) + \text{Re} \ \{f'(\xi_1) \cdot (z - z_0)\} + i \cdot \text{Im} \ \{f'(\xi_2) \cdot (z - z_0)\}.$$

f' denotes the gradient of f.

In the following we mostly use an immediate conclusion:

Theorem 6:

Let z_1, z_2 and $\tilde{z} \in V_n\mathbb{C}$. The set Z is defined by

$$Z := \cap \{Z \in IV_n\mathbb{C} \mid z_1 \in Z \text{ and } z_2 \in Z\}.$$

If $G \in PV_n\mathbb{C}$, $\tilde{z} \cup Z \subseteq G$ and $f: G \longrightarrow V_n\mathbb{C}$ holomorph then

$$f(Z) \subseteq f(\tilde{z}) + f'(\tilde{z} \cup Z) \cdot (Z - \tilde{z}),$$

where the Jacobian matrix f' is defined by

$$f'(Z) := \cap \{Y \in IM_n\mathbb{C} \mid \inf(Y) \leq f'(z) \leq \sup(Y) \text{ for all } z \in Z\}.$$

Now the central theorem for the inclusion of nonlinear complex systems of equations can be derived:

Theorem 7:

Let $G \in PV_n\mathbb{C}$, $R \in M_n\mathbb{C}$, $\tilde{z} \in V_n\mathbb{C}$ and $f: G \longrightarrow V_n\mathbb{C}$ be a holomorphic function. If then for some $Z \in IV_n\mathbb{C}$ with $\tilde{z} \cup Z \subseteq G$

$$\tilde{z} - R \cdot f(\tilde{z}) + \{E - R \cdot f'(\tilde{z} \cup Z)\} \cdot (Z - \tilde{z}) \subseteq Z$$

then there exists one and only one $\hat{z} \in Z$ with $f(\hat{z}) = 0$ and the matrices R and all $M \in f'(\tilde{z} \cup Z)$ are nonsingular.

3. Inclusion of the complex algebraic eigenvalue problem

The eigenvalue problem can be considered as the special case of the general problem of nonlinear systems. The task is to determine the unknowns $\lambda \in \mathbb{C}$ and $x \in V_n\mathbb{C}$ by the system

$$A \cdot x - \lambda x = 0$$

$$e_k^T \cdot x - \xi = 0$$

with $A \in M_n\mathbb{C}$ and $\xi \neq 0$. Due to the special structure of these nonlinear systems, a more detailed statement can be given for the inclusion of the eigenvalues and the corresponding eigenvectors.

Theorem 8:

Let $A \in M_n\mathbb{C}$, $R \in M_{n+1}\mathbb{C}$, $\tilde{x} \in V_n\mathbb{C}$, $\tilde{\lambda} \in \mathbb{C}$ and $0 \neq \xi \in \mathbb{C}$.

For $Y \in IV_n\mathbb{C}$ and $M \in I\mathbb{C}$ the function $G: IV_{n+1}\mathbb{C} \longrightarrow PV_{n+1}\mathbb{C}$ is defined by

$$G \begin{bmatrix} Y \\ M \end{bmatrix} := \begin{bmatrix} \tilde{x} \\ \tilde{\lambda} \end{bmatrix} - R \cdot \begin{bmatrix} A \cdot \tilde{x} & - \tilde{\lambda} \cdot \tilde{x} \\ e_k^T \cdot \tilde{x} & - \xi \end{bmatrix} +$$

$$+ (E_{n+1} - R \cdot \begin{bmatrix} A - (\tilde{\lambda} \cup M) \cdot E & - (\tilde{x} \cup Y) \\ e_k^T & 0 \end{bmatrix}) \cdot \begin{bmatrix} Y - \tilde{x} \\ M - \tilde{\lambda} \end{bmatrix} \cdot$$

If for some $X \in IV_n\mathbb{C}$ and some $\Lambda \in I\mathbb{C}$

$$G \begin{bmatrix} X \\ \Lambda \end{bmatrix} \subseteq \begin{bmatrix} X \\ \Lambda \end{bmatrix}$$

then there exists one and only one eigenvalue/eigenvector pair $(\hat{\lambda}, \hat{x})$ with $\hat{\lambda} \in \Lambda$ and $\hat{x} \in X$.

In order to obtain faster and more accurate inclusions it is necessary to reduce the diameter of the Jacobian matrix

$$J(X, \Lambda) := \begin{bmatrix} A - (\tilde{\lambda} \cup \Lambda) \cdot E & - (\tilde{x} \cup X) \\ e_k^T & 0 \end{bmatrix} \cdot$$

The interval matrix $S(X) \in IM_{n+1}\mathbb{C}$ is defined as

$$S(X) := \begin{bmatrix} A - \tilde{\lambda} \cdot E & -X \\ e_k^T & 0 \end{bmatrix}.$$

Corresponding to the function G above (Theorem 8) the function

$$H \begin{bmatrix} X \\ \Lambda \end{bmatrix} := \begin{bmatrix} \tilde{x} \\ \tilde{\lambda} \end{bmatrix} - R \cdot \begin{bmatrix} A \cdot \tilde{x} & - \tilde{\lambda} \cdot \tilde{x} \\ e_k^T \cdot \tilde{x} & - \xi \end{bmatrix} +$$

$$+ (E_{n+1} - R \cdot S(X)) \cdot \begin{bmatrix} X - \tilde{x} \\ \Lambda - \tilde{\lambda} \end{bmatrix}$$

can be constructed.

Because of $S(X) \subseteq J(X, \Lambda)$ it is evidently

$$H \begin{bmatrix} X \\ \Lambda \end{bmatrix} \subseteq G \begin{bmatrix} X \\ \Lambda \end{bmatrix}.$$

The following theorem proves the existence of the eigenvalue/eigenvector pair also under these weaker assumptions. Since $G(T) \, \mathcal{E} \, T$ does not follow from $H(T) \, \mathcal{E} \, T$, Theorem 9 has to be proven unaided by Theorem 7.

Theorem 9:

Let $A \in M_n\mathbb{C}$, $R \in M_{n+1}\mathbb{C}$, $\tilde{x} \in V_n\mathbb{C}$, $\tilde{\lambda} \in \mathbb{C}$ and $0 \neq \xi \in \mathbb{C}$. For $Y \in IV_n\mathbb{C}$ and $M \in I\mathbb{C}$ the function $H: IV_{n+1}\mathbb{C} \longrightarrow PV_{n+1}\mathbb{C}$ is defined by

$$H \begin{bmatrix} Y \\ M \end{bmatrix} := \begin{bmatrix} \tilde{x} \\ \tilde{\lambda} \end{bmatrix} - R \cdot \begin{bmatrix} A \cdot \tilde{x} & - \tilde{\lambda} \cdot \tilde{x} \\ e_k^T \cdot \tilde{x} & - \xi \end{bmatrix} +$$

$$+ \begin{bmatrix} E_{n+1} - R \cdot \begin{bmatrix} A - \tilde{\lambda} \cdot E & -Y \\ e_k^T & 0 \end{bmatrix} \end{bmatrix} \cdot \begin{bmatrix} Y - \tilde{x} \\ M - \tilde{\lambda} \end{bmatrix}.$$

If for some $X \in IV_n\mathbb{C}$ and some $\Lambda \in I\mathbb{C}$

$$H \begin{bmatrix} X \\ \Lambda \end{bmatrix} \mathcal{E} \begin{bmatrix} X \\ \Lambda \end{bmatrix}$$

then there exists at least one eigenvalue $\hat{\lambda} \in \Lambda$ and one eigenvetor $\hat{x} \in X$ with $A \cdot \hat{x} = \hat{\lambda} \cdot \hat{x}$. The eigenvalue $\hat{\lambda}$ possesses the multiplicity one.

<u>Proof:</u> We use the following notation:

$$DZ := \begin{bmatrix} \tilde{x} \\ \tilde{\lambda} \end{bmatrix} - R \cdot \begin{bmatrix} A \cdot \tilde{x} & - \tilde{\lambda} \cdot \tilde{x} \\ e_k^T \cdot \tilde{x} & - \xi \end{bmatrix}.$$

We construct the function $h := V_{n+1}\mathbb{C} \longrightarrow V_{n+1}\mathbb{C}$ in the following way

$$h \begin{bmatrix} x \\ \lambda \end{bmatrix} := DZ + \begin{bmatrix} E_{n+1} - R \cdot \begin{bmatrix} A - \tilde{\lambda}\cdot E & - x \\ e_k^T & 0 \end{bmatrix} \end{bmatrix} \cdot \begin{bmatrix} x - \tilde{x} \\ \lambda - \tilde{\lambda} \end{bmatrix}$$

If $x \in X$ and $\lambda \in \Lambda$ it holds evidently

$$h(t) \in H(T) \ ? \ T \quad \text{with } t := (x, \lambda)^T \text{ and } T := (X, \Lambda)^T.$$

Thus the function fulfils the conditions of Theorem 2, thus h has at least one fixed-point in T.

Since

$$h \begin{bmatrix} x \\ \lambda \end{bmatrix} = \begin{bmatrix} \tilde{x} \\ \tilde{\lambda} \end{bmatrix} - R \cdot \begin{bmatrix} A \cdot \tilde{x} & - \tilde{\lambda} \cdot \tilde{x} \\ e_k^T \cdot \tilde{x} & - \xi \end{bmatrix} + \begin{bmatrix} x - \tilde{x} \\ \lambda - \tilde{\lambda} \end{bmatrix} -$$

$$- R \cdot \begin{bmatrix} A \cdot x - A \cdot \tilde{x} - \tilde{\lambda} \cdot x + \tilde{\lambda} \cdot \tilde{x} - \lambda \cdot x + \tilde{\lambda} \cdot x \\ e_k^T \cdot (x - \tilde{x}) \end{bmatrix} =$$

$$= \begin{bmatrix} x \\ \lambda \end{bmatrix} - R \cdot \begin{bmatrix} A \cdot x & - \lambda \cdot x \\ e_k^T \cdot x & - \xi \end{bmatrix}$$

and because of the existence of a fixed-point $(\hat{x}, \hat{\lambda})$ with $\hat{x} \in X$ and $\hat{\lambda} \in \Lambda$

$$h \begin{bmatrix} \hat{x} \\ \hat{\lambda} \end{bmatrix} = \begin{bmatrix} \hat{x} \\ \hat{\lambda} \end{bmatrix} - R \cdot \begin{bmatrix} A \cdot \hat{x} & - \hat{\lambda} \cdot \hat{x} \\ e_k^T \cdot \hat{x} & - \xi \end{bmatrix} = \begin{bmatrix} \hat{x} \\ \hat{\lambda} \end{bmatrix},$$

it follows that

$$R \cdot \begin{bmatrix} A \cdot \hat{x} & - \hat{\lambda} \cdot \hat{x} \\ e_k^T \cdot \hat{x} & - \xi \end{bmatrix} = 0.$$

With $Y := T - t$ it holds $DZ + C \cdot Y \stackrel{?}{=} Y$ for every

$$C \in \begin{bmatrix} E_{n+1} - R \cdot \begin{bmatrix} A - \tilde{\lambda} \cdot E & - X \\ e_k^T & 0 \end{bmatrix} \end{bmatrix}.$$

From Theorem 4 it follows that R is nonsingular. Thus we have

$$\begin{bmatrix} A \cdot \hat{x} & - \hat{\lambda} \cdot \hat{x} \\ e_k^T \cdot \hat{x} & - \xi \end{bmatrix} = 0 \quad \text{i.e.} \quad \begin{matrix} A \cdot \hat{x} = \hat{\lambda} \cdot \hat{x} \\ e_k^T \cdot \hat{x} = \xi \neq 0 \end{matrix}.$$

We change the function h to the function $h^* : V_{n+1}\mathbb{C} \longrightarrow V_{n+1}\mathbb{C}$ in the following way

$$h^* \begin{bmatrix} x \\ \lambda \end{bmatrix} = DZ + \begin{bmatrix} E_{n+1} - R \cdot \begin{bmatrix} A - \lambda \cdot E & - \tilde{x} \\ e_k^T & 0 \end{bmatrix} \end{bmatrix} \cdot \begin{bmatrix} x - \tilde{x} \\ \lambda - \tilde{\lambda} \end{bmatrix}.$$

Then it holds that

$$h^* \begin{bmatrix} x \\ \lambda \end{bmatrix} = \begin{bmatrix} \tilde{x} \\ \tilde{\lambda} \end{bmatrix} - R \cdot \begin{bmatrix} A \cdot \tilde{x} & - \tilde{\lambda} \cdot \tilde{x} \\ e_k^T \cdot \tilde{x} & - \xi \end{bmatrix} + \begin{bmatrix} x - \tilde{x} \\ \lambda - \tilde{\lambda} \end{bmatrix} -$$

$$- R \cdot \begin{bmatrix} A \cdot x - A \cdot \tilde{x} - \lambda \cdot x + \lambda \cdot \tilde{x} - \lambda \cdot \tilde{x} + \tilde{\lambda} \cdot \tilde{x} \\ e_k^T \cdot (x - \tilde{x}) \end{bmatrix} =$$

$$= \begin{bmatrix} x \\ \lambda \end{bmatrix} - R \cdot \begin{bmatrix} A \cdot x & - \lambda \cdot x \\ e_k^T \cdot x & - \xi \end{bmatrix} = h \begin{bmatrix} x \\ \lambda \end{bmatrix}.$$

With $t \in T$ and $Y := T - t$, it also holds that $h^*(t) \in H(T) \overset{\sim}{\subset} T$ and $DZ + C^* \cdot Y \overset{\sim}{\subset} Y$ for every

$$
C^* \in \left[E_{n+1} - R \cdot \begin{bmatrix} A - \Lambda \cdot E & -\tilde{x} \\ e_k^T & 0 \end{bmatrix} \right].
$$

From Theorem 3 it follows that $\rho(C^*) < 1$ and by Theorem 4 we have that

$$
\begin{bmatrix} A - \lambda \cdot E & -\tilde{x} \\ e_k^T & 0 \end{bmatrix} \text{ is nonsingular for every } \lambda \in \Lambda.
$$

In particular the matrix

$$
W := \begin{bmatrix} A - \hat{\lambda} \cdot E & -\tilde{x} \\ e_k^T & 0 \end{bmatrix} \text{ is nonsingular, i.e. rank } (W) = n+1.
$$

From the assumption, that $\hat{\lambda}$ has a multiplicity greater than 1, it follows rank $(A - \hat{\lambda} E) \leq n - 2$. With the increase of this matrix by one column and one row, the rank of the matrix could run up to n at most. This is a contradiction to the proven result that rank $(W) = n + 1$. Thus we obtain that $\hat{\lambda}$ has the multiplicity one.

Probably it holds in Theorem 9 that there is one and only one eigenvalue/ eigenvector pair, but the uniquness can not yet be proved. There are also other approaches to reduce the diameter of the Jacobian matrix. The result, however, was the same as that shown above.

In practical computation it is often advantagous (e.g. see linear systems) to calculate not an inclusion of the solution, but a good approximation of the solution and then to include the defect, the difference between the exact solution and the approximation.

With $H(T) \overset{\sim}{\subset} T$ it also holds

$$
H(T) - \begin{bmatrix} \tilde{x} \\ \tilde{\lambda} \end{bmatrix} \overset{\sim}{\subset} T - \begin{bmatrix} \tilde{x} \\ \tilde{\lambda} \end{bmatrix}.
$$

Let us define

$$U := T - \begin{bmatrix} \tilde{x} \\ \tilde{\lambda} \end{bmatrix} \qquad \text{and} \qquad F(U) := H(T) - \begin{bmatrix} \tilde{x} \\ \tilde{\lambda} \end{bmatrix},$$

then the matrix A possesses under the conditions of Theorem 9 at least one eigenvalue/eigenvector pair in $U + (\tilde{x}, \tilde{\lambda})^T$.

We will summarize these results in

Theorem 10:

Let $A \in M_n\mathbb{C}$, $R \in M_{n+1}\mathbb{C}$, $\tilde{x} \in V_n\mathbb{C}$, $\tilde{\lambda} \in \mathbb{C}$ and $0 \neq \xi \in \mathbb{C}$.

For $Y \in IV_n\mathbb{C}$ and $M \in I\mathbb{C}$ the function $F : IV_{n+1}\mathbb{C} \longrightarrow PV_{n+1}\mathbb{C}$ is defined by

$$F\begin{bmatrix} Y \\ M \end{bmatrix} := - R \cdot \begin{bmatrix} A \cdot \tilde{x} & - \tilde{\lambda} \cdot \tilde{x} \\ e_k^T \cdot \tilde{x} & - \xi \end{bmatrix}$$

$$+ \begin{bmatrix} E_{n+1} - R \cdot \begin{bmatrix} A - \tilde{\lambda} \cdot E & - \tilde{x} - Y \\ e_k^T & 0 \end{bmatrix} \end{bmatrix} \cdot \begin{bmatrix} Y \\ M \end{bmatrix}.$$

If for some $U \in IV_{n+1}\mathbb{C}$

$$F(U) \subseteq U \text{ with } U = \begin{bmatrix} X \\ \Lambda \end{bmatrix}.$$

there exists at least one eigenvalue $\hat{\lambda} \in \tilde{\lambda} + \Lambda$ and at least one eigenvector $\hat{x} \in \tilde{x} + X$ and it holds that $A \cdot \hat{x} = \hat{\lambda} \cdot \hat{x}$. The eigenvalue $\hat{\lambda}$ possesses the multiplicity one.

The proof runs in a similar way as in Theorem 10. In particular, we obtain

$$e_k^T \cdot (\tilde{x} + \hat{x}) = \xi.$$

If the kth component of x is scaled so that $e_k^T \cdot \tilde{x} = \xi$, we receive
$e_k^T \cdot \hat{x} = 0$. Thus in Theorem 10, the vector

$$\begin{bmatrix} A \cdot \tilde{x} & - \tilde{\lambda} \cdot \tilde{x} \\ e_k^T \cdot \tilde{x} & - \xi \end{bmatrix} \text{ can be replaced by } \begin{bmatrix} A \cdot \tilde{x} & - \tilde{\lambda} \cdot \tilde{x} \\ & 0 \end{bmatrix}.$$

We are now able to consider the implementation of Theorem 10 on a computer:

Theorem 11:

Let $A \in M_n\mathbb{C}$, $R \in M_{n+1}\mathbb{C}$, $\tilde{x} \in V_n\mathbb{C}$ and $\tilde{\lambda} \in \mathbb{C}$. For $Y \in IV_n\mathbb{C}$ and $M \in I\mathbb{C}$ the function $F_\diamondsuit : IV_{n+1}\mathbb{C} \longrightarrow IV_{n+1}\mathbb{C}$ is defined by

$$F_\diamondsuit \begin{bmatrix} Y \\ M \end{bmatrix} := \diamondsuit R \diamondsuit \diamondsuit \begin{bmatrix} A \cdot \tilde{x} & - \tilde{\lambda} \cdot \tilde{x} \\ & 0 \end{bmatrix}$$

$$\diamondsuit \diamondsuit \begin{bmatrix} E_{n+1} - R \cdot \begin{bmatrix} A - \tilde{\lambda} \cdot E & - \tilde{x} - Y \\ e_k^T & 0 \end{bmatrix} \end{bmatrix} \diamondsuit \begin{bmatrix} Y \\ M \end{bmatrix}.$$

If for some $U \in IV_{n+1}\mathbb{C}$

$$F_\diamondsuit(U) \subseteq U \text{ with } U = \begin{bmatrix} X \\ \Lambda \end{bmatrix},$$

then there exists at least one eigenvalue $\hat{\lambda} \in \tilde{\lambda} \diamondsuit \Lambda$ and at least one eigenvector $\hat{x} \in \tilde{x} \diamondsuit X$ with $A \cdot \hat{x} = \hat{\lambda} \cdot \hat{x}$. The eigenvalue $\hat{\lambda}$ possesses the multiplicity one.

Proof: It is evident that $F(U) \subseteq F_\diamondsuit(U)$. Thus with $F_\diamondsuit(U) \subseteq U$, it also holds that $F(U) \subseteq U$ and therefore Theorem 10 can be applied.

4. The Inclusion Algorithm

Some considerations are necessary to derive an efficient algorithm from the mathematical foundations. Nothing was said in the previous theorems about the matrix R. R can be chosen as approximate inverse of the matrix

$$\begin{bmatrix} A - \tilde{\lambda} \cdot E & -\tilde{x} \\ e_k^T & 0 \end{bmatrix}.$$

Thus the kth row of R has the form: $(0\ 0\ \ldots\ 0\ 1)$ and with $\tilde{x}_k = \xi$, we have

$$f\left(\begin{bmatrix} x_1 \\ \vdots \\ x_k \\ \vdots \\ x_n \\ \lambda \end{bmatrix}\right) = \begin{bmatrix} y_1 \\ \vdots \\ 0 \\ \vdots \\ y_n \\ m \end{bmatrix}.$$

The problem of the dimension n+1 can thus be transformed to such one of the dimension n, and thereby the function F in Theorem 10 receives the structure

$$F(Y) := -R \cdot ((A - \tilde{\lambda} \cdot E) \cdot \tilde{x}) + (E - R \cdot B(X)) \cdot Y \quad \text{with}$$

$$B(X) := ((A - \tilde{\lambda} \cdot E)_1, \ldots (A - \tilde{\lambda} \cdot E)_{k-1}, -\tilde{x} - X, (A - \tilde{\lambda} \cdot E)_{k+1}, \ldots (A - \tilde{\lambda} \cdot E)_n)$$

and

$$R \approx ((A - \tilde{\lambda} \cdot E)_1, \ldots (A - \tilde{\lambda} \cdot E)_{k-1}, -\tilde{x}, (A - \tilde{\lambda} \cdot E)_{k+1}, \ldots (A - \tilde{\lambda} \cdot E)_n)^{-1}.$$

The inclusion condition is transformed to

$$F(Y) \subseteq Y \quad \text{with} \quad Y = \begin{bmatrix} x_1 \\ \vdots \\ \wedge \\ \vdots \\ x_n \end{bmatrix} \in IV_n \; \mathbb{C}.$$

It is attempted to achieve this condition by means of a residuum interval iteration which is defined by the function F:

Starting with the interval $Y^0 := -R \; \diamondsuit \; \diamondsuit \; ((A - \tilde{\lambda} \cdot E) \cdot \tilde{x})$ the intervals Y^l, $l \geq 1$, are computed by

$$Y^{l+1} := Y^0 \; \diamondsuit \; \diamondsuit \; (E - R \cdot B(X^l)) \; \diamondsuit \; Y^l, \; l \geq 0$$

until for some l $\quad Y^{l+1} \subseteq Y^l$ is reached. The intervals $X^l = (x_i^l)$ has the structure

$$x_i^l := \begin{cases} yl & i \neq k \\ 0 & i = k \end{cases} \quad \text{for} \quad .$$

If an interval already lies near by a fixed-point, it is a usual method to catch this point in the interval by blowing up the interval slightly. Let X denote the complex interval

$$X = ([x.re.inf \; , \; x.re.sup] \; , \; [x.im.inf \; , \; x.im.sup])$$

and F_{re} is defined by

$$F_{re} := 0.1 \cdot (x.re.sup - x.re.inf).$$

If minpos $:= 1.0E-emax$ and $\gamma := 1.0E-(m-1)$ where m denotes the mantissa length, then the interval blow up is characterized by $XB := X \circ \sigma$. The real components of XB are computed by

$$xb.re.inf := x.re.inf - eps \cdot (F_{re} + \gamma \cdot |x.re.inf|) - minpos$$
$$xb.re.sup := x.re.sup + eps \cdot (F_{re} + \gamma \cdot |x.re.sup|) + minpos.$$

The imaginary components are computed analogously. Such an interval blow up is embedded in the interval iteration. In the first steps, eps is defined by eps := 1. If after 3 iterations no inclusion is occured then eps is increased by 5 and so on.

Suppose there is already computed an approximation pair $(\tilde{\lambda}, \tilde{x})$ for the eigenvalue/eigenvector by an usual algorithm. In many cases the approximation is not good enough to achieve the described inclusion. The first approximation can then be improved by the following Newton method:

$$y_i^0 := \tilde{x}_i \quad \text{for } i \neq k \quad \text{and} \quad y_k^0 := \tilde{\lambda}$$
$$y^{l+1} := y^l - R \cdot ((A - \lambda^l \cdot E) \cdot x^l) \quad \text{with}$$

$$\lambda^l = y_k^l, \quad x_i^l = y_i^l \quad \text{for } i \neq k \quad \text{and} \quad x_k^l = \tilde{x}_k.$$

Several methods to compute the approximate inverse R have been tested, such as LU-decomposition and so on. The Gauß-Jordan-algorithm has turned out to be the best method in comparison to the effort.

Finally, let us consider how to get a good approximation for the eigenvalue/ eigenvector pair. In the general complex case, the LR-algorithm described in [10] was the best under the tested methods. The LR-algorithm for the complex matrix A is characterized by the following 5 steps:

1) Isolation of the eigenvalues by matrix transformations and computation of a balanced matrix B with better condition.
2) Transformation of B to the Hessenberg form H.
3) Computation of the eigenvalues and eigenvectors of H by means of the modified LR-algorithm.
4) Backtransformation of the eigenvectors of H to the eigenvectors of B.
5) Backtransformation of the eigenvectors of B to the eigenvectors of A.

We are now able to summarize the algorithm completely:

Inclusion of a separate eigenvalue and the corresponding eigenvectors of any complex matrix A.

1) Compute a floating-point approximation of the complex eigenvalue $\tilde{\lambda}$ and of the corresponding eigenvector $\tilde{x}$ using the LR-algorithm.

$$\tilde{x}_k := \max_{1 \leq i \leq n} \tilde{x}_i \ .$$

2) Compute by floating-point arithmetic the approximate inverse

$$R \approx ((A - \tilde{\lambda} \cdot E)_1, \ldots (A - \tilde{\lambda} \cdot E)_{k-1}, -\tilde{x}, (A - \tilde{\lambda} \cdot E)_{k+1}, \ldots (A - \tilde{\lambda} \cdot E)_n)^{-1}.$$

using the Gauß-Jordan-algorithm.

if 'R singular' then ready := false; stop {No inclusion}.

3) Improve $\tilde{\lambda}$ and $\tilde{x}$ using floating-point arithmetic

$l := 0;$
$x^0 := \tilde{x}; \qquad \lambda^0 := \tilde{\lambda};$
repeat
   $\Delta := 0;$
   $dx^{l+1} := R \boxdot \square((A - \lambda^l \cdot E) \cdot x^l);$
   $d\lambda^{l+1} := dx_k^{l+1}; \qquad dx_k^{l+1} := 0;$
   $\lambda^{l+1} := \lambda^l - d\lambda^{l+1};$
   $x^{l+1} := x^l - dx^{l+1};$
   for i := 1 to n do $\Delta := \max (\Delta , |x_i^{l+1} - x_i^l| / |x_i^l|);$
   $\Delta := \max (\Delta , |\lambda^{l+1} - \lambda^l| / |\lambda^l|);$
   $l := l + 1;$
until $(\Delta \leq \mu)$ or $(l \geq 50)$ or 'iteration divergent';
$\{\mu$ is a constant depending of the used floating-point system$\}$
if 'iteration divergent' then
   begin $\lambda := \tilde{\lambda};$ $x := \tilde{x}$ end
else
   begin $\lambda := \lambda^l;$ $x := x^l$ end

4) Interval computation of the defect

$$Z := - R \diamondsuit \diamondsuit((A - \lambda \cdot E) \cdot x)$$

5) Interval iteration

$$1 := -1; \quad Y^0 := Z;$$
$$\text{ready} := \text{false};$$
$\underline{\text{repeat}}$
$\quad 1 := 1 + 1;$
$\quad YB^l := Y^l \ \sigma; \qquad\qquad \{\text{interval blow up}\}$
$\qquad XB^l := YB^l; \quad XB_k^l := 0;$
$\qquad Y^{l+1} := Z \otimes \diamond (E - R \cdot B(XB^l)) \otimes YB^l; \qquad \text{with}$
$\qquad B(XB^l) := ((A - \lambda E)_1, \ldots (A - \lambda E)_{k-1}, -x - XB^l, (A - \tilde{\lambda} E)_{k+1}, \ldots (A - \lambda E)_n);$
$\qquad \text{ready} := (Y^{l+1} \subset YB^l);$
$\underline{\text{until}}$ ready $\underline{\text{or}}$ diam(Y^{l+1}) too large $\underline{\text{or}}$ (1>q).

6) Result

$\underline{\text{if}}$ ready $\underline{\text{then}}$ {inclusion achieved}
$\quad \underline{\text{begin}}$
$\qquad \wedge := \lambda \otimes Y_k^{l+1}; \quad Y_k^{l+1} := 0;$
$\qquad X := x \otimes Y^{l+1};$
$\quad \underline{\text{end}}$ { it has been verified that there exists an eigenvalue $\hat{\lambda} \ \epsilon \ \wedge$ and
$\qquad$ an eigenvector $\hat{x} \ \epsilon \ X$ with $A \cdot \hat{x} = \hat{\lambda} \cdot \hat{x}$ and $\hat{\lambda}$ has the multiplicity
$\qquad$ one }
$\underline{\text{else}}$ { no verification }.

The algorithm presented here has been implemented in PASCAL-SC on the SAM 68K. The method is available as a subroutine in other programs. The user has to define the following objects in his own program:

```
CONST    maxdim   = 15;
TYPE     dimtype  = 1 .. maxdim;
         rvector  = ARRAY[dimtype] OF real;
         cvector  = ARRAY[dimtype] OF complex;
         cmatrix  = ARRAY[dimtype] OF cvector;
         civector = ARRAY[dimtype] OF cinterval;
```

and the head of the procedure:

```
    PROCEDURE CEIG (dim : dimtype; VAR a : cmatrix; lambda : complex;
                    x : cvector; VAR iclambda : cinterval;
                    VAR icx : civector; VAR error : integer );
EXTERNAL;
```

Example: The Hilbert matrix of the dimension n is defined by

$$H_n := (h_{ij}) \quad \text{with} \quad h_{ij} = \frac{1}{i+j-1}, \qquad i, j = 1(1)n.$$

Let us consider the complex eigenvalue problem

$$(k_n \cdot H_n , k_n \cdot H_n) \cdot x = \lambda \cdot x$$

where k_n is the least common multiple of $(1, \ldots, 2n-1)$.

The problem was solved by the described algorithm implemented in PASCAL-SC on a SAM 68K computer (13 decimal digits). In comparison, the problem has been calculated by NAG-library routine on a Siemens 7881 ($\approx$ 17 decimal digits). For a dimension of 11, we give the results obtained for the smallest eigenvalue and the corresponding eigenvector:

a) SAM 68K

11th eigenvalue:

$$\left(\begin{bmatrix} 7.899160434834E\text{-}07, & 7.899160434835E\text{-}07 \\ 7.899160434834E\text{-}07, & 7.899160434835E\text{-}07 \end{bmatrix} \right)$$

11th eigenvector:

$$\left(\begin{bmatrix} -4.918154923946E\text{-}07, & -4.918154923945E\text{-}07 \\ -4.918154923946E\text{-}07, & -4.918154923945E\text{-}07 \end{bmatrix} \right)$$

$$\left(\begin{bmatrix} 5.240523388557E\text{-}05, & 5.240523388558E\text{-}05 \\ 5.240523388557E\text{-}05, & 5.240523388558E\text{-}05 \end{bmatrix} \right)$$

$$\left(\begin{bmatrix} -1.378821269177E\text{-}03, & -1.378821269176E\text{-}03 \\ -1.378821269177E\text{-}03, & -1.378821269176E\text{-}03 \end{bmatrix} \right)$$

$$\left(\begin{bmatrix} 1.559486813473\text{E-}02, & 1.559486813474\text{E-}02 \\ 1.559486813473\text{E-}02, & 1.559486813474\text{E-}02 \end{bmatrix} \right)$$

$$\left(\begin{bmatrix} -9.380533307375\text{E-}02, & -9.380533307374\text{E-}02 \\ -9.380533307375\text{E-}02, & -9.380533307374\text{E-}02 \end{bmatrix} \right)$$

$$\left(\begin{bmatrix} 3.325111516091\text{E-}01, & 3.325111516092\text{E-}01 \\ 3.325111516091\text{E-}01, & 3.325111516092\text{E-}01 \end{bmatrix} \right)$$

$$\left(\begin{bmatrix} -7.290786316947\text{E-}01, & -7.290786316946\text{E-}01 \\ -7.290786316947\text{E-}01, & -7.290786316946\text{E-}01 \end{bmatrix} \right)$$

$$\left(\begin{bmatrix} 1.000000000000\text{E+}00, & 1.000000000000\text{E+}00 \\ 1.000000000000\text{E+}00, & 1.000000000000\text{E+}00 \end{bmatrix} \right)$$

$$\left(\begin{bmatrix} -8.350795249985\text{E-}01, & -8.350795249984\text{E-}01 \\ -8.350795249985\text{E-}01, & -8.350795249984\text{E-}01 \end{bmatrix} \right)$$

$$\left(\begin{bmatrix} 3.881865349939\text{E-}01, & 3.881865349940\text{E-}01 \\ 3.881865349939\text{E-}01, & 3.881865349940\text{E-}01 \end{bmatrix} \right)$$

$$\left(\begin{bmatrix} -7.700251644188\text{E-}02, & -7.700251644187\text{E-}02 \\ -7.700251644188\text{E-}02, & -7.700251644187\text{E-}02 \end{bmatrix} \right)$$

b) Siemens 7881:

11th eigenvalue:

(0.79271312073844D-06,   0.79271312073844D-06)

11th eigenvector:

(-0.49181878060388D-06,   -0.49181878060388D-06)
(0.52405559539299D-04,   0.52405559539299D-04)
(-0.13788283609873D-02,   -0.13788283609873D-02)
(0.15594929725873D-01,   0.15594929725873D-01)
(-0.93805595913356D-01,   -0.93805595913356D-01)
(0.33251173562567D+00,   0.33251173562567D+00)
(-0.72907923255878D+00,   -0.72907923255878D+00)
(0.10000000000000D+01,   0.10000000000000D+01)
(-0.83507891793239D+00,   -0.83507891793239D+00)
(0.38818600364914D+00,   0.38818600364914D+00)
(-0.77002367294144D-01,   -0.77002367294144D-01)

References

[1] Alefeld, G., Herzberger, J.: Einführung in die Intervallrechnung. Bibliographisches Institut Mannheim, Wien Zürich (1974).

[2] Bohlender, G.: Genaue Berechnung mehrfacher Summen, Produkte und Wurzeln von Gleitkommazahlen und allgemeine Arithmetik in Höheren Programmiersprachen. Dissertation, Universität Karlsruhe (1978).

[3] Grüner, K.: Allgemeine Rechnerarithmetik und deren Implementierung. Dissertation, Universität Karlsruhe (1979).

[4] Kulisch, U., Miranker, W. L.: Computer Arithmetic in Theory and Practice. Academic Press (1981).

[5] Kulisch, U., Miranker, W. L. (Eds.): A New Approach to Scientific Computation. Academic Press (1983).

[6] Wilkinson, J. H.: The Algebraic Eigenvalue Problem. Clarendon Press, Oxford (1965).

[7] Wilkinson, J. H., Reinsch, C.: Handbook for automatic computation. Volume 2: Linear Algebra. Springer-Verlag, Berlin (1971).

Techniques for Generating Accurate Eigensolutions in ADA
DIAMOND deliverable D 3-1 (part)

V. Moynihan[t]

Numerical Algorithms Group Ltd, Wilkinson House,
Jordan Hill Road, Oxford, United Kingdom, OX2 8DR

Abstract

We consider the use of accurate arithmetic and interval primitives to both increase and guarantee the accuracy of eigensolutions. In most cases the accuracy delivered is such that no other machine number lies between the computed result and the true value.

Keywords: *Eigenvalues, eigenvectors, iterative refinement, fixed point theorems, optimal accuracy, generics.*

1. Introduction

This paper details a strategy for obtaining eigensolutions of matrices with high accuracy, efficiently. This efficiency results from the judicious use of accurate arithmetic and an economical algorithm for performing the residual correction and enclosure of the solution. The accurate arithmetic used in the implementation was obtained from an Ada package called GENERIC_SCIENTIFIC_COMPUTATION [1].

In the algorithms described here error control is performed automatically. The user is not left with the problem of deciding whether the results are valid. This usually requires reruns with slightly modified data, or re-solving the problem in greater precision **throughout** the algorithm, this may well be wasteful. Thus the process of obtaining results from data has been de-skilled allowing the scientist to have confidence in his results without recourse to a numerical analyst. It is important to state that the problem solved is that which is represented in the number set used by the machine. In ill-conditioned problems, it is well known that machine numbers may not represent the original problem with enough precision to solve the system accurately. In those cases, the corresponding interval problem is solved.

2. Method

We begin by discussing two techniques, iterative refinement and accurate inclusion of numerical results. We will present an economic iterative refinement algorithm, and show how the inclusion of an eigensolution can be made more economic using work from the iterative refinement algorithm.

2.1. Derivation of Economic Iterative Refinement Algorithm

Following Dongarra [5], who deals with the real case, the following algorithm valid for the complex case can be derived. The real case is closely related, there being an additional series of rotations to first reduce a quasi-triangular matrix to strictly triangular form.

If λ, x is an eigenpair of a matrix A and $\lambda+\mu$, $x+\bar{y}$ is a neighbouring pair then

$$A(x+\bar{y}) = (\lambda+\mu)(x+\bar{y}) \tag{1}$$

is satisfied exactly. Assume x is normalised i.e. $\|x\|_\infty = 1$, where $x_s = 1$. This implies a degree of arbitrariness in $\bar{y}$, consider the 2 by 2 system where $x_1 = 1$.

$$\begin{bmatrix} a_{1,1} & a_{1,2} \\ a_{2,1} & a_{2,2} \end{bmatrix} \begin{bmatrix} 1+\bar{y}_1 \\ x_2+\bar{y}_2 \end{bmatrix} = (\lambda+\mu) \begin{bmatrix} 1+\bar{y}_1 \\ x_2+\bar{y}_2 \end{bmatrix}$$

---

[t] Current address: Computer Sciences Company Ltd, Computer Sciences House, Brunel Way, Slough, Berkshire, United Kingdom.

If A, λ and μ are fixed then we have two equations in three unknowns. Thus, we can set $\bar{y}_{s=1}$ to be zero.

Rearrange (1)

$$(A-\lambda E)\bar{y} - \mu x = \lambda x - Ax + \mu\bar{y}. \tag{2}$$

Introduce $y' = (\bar{y}_1 \ \cdots \ \bar{y}_{s-1}, \mu, \bar{y}_{s+1} \ \cdots \ \bar{y}_n)$ and

$$B = \begin{bmatrix} a_{1,1}-\lambda & a_{1,2} & \cdots & a_{1,s-1} & -x_1 & a_{1,s+1} & a_{1,n} \\ \cdot & & & & & & \\ \cdot & & & & & & \\ \cdot & & & & & & \\ a_{n,1} & a_{n,2} & \cdots & a_{n,s-1} & -x_n & a_{n,s+1} & a_{n,n}-\lambda \end{bmatrix}$$

Therefore

$$By = (A-\lambda E)\bar{y} - \mu x \tag{3}$$

and

$$y, \ \bar{y} = \mu\bar{y}$$

By letting $r = \lambda x - Ax$, we can then say

$$By = r + y_s\bar{y}. \tag{4}$$

Another way to view equation (3) is to consider the matrix $\bar{B}$ as a $(n+1)\times(n+1)$ matrix of the form

$$\bar{B} = \begin{bmatrix} A-\lambda E & -x \\ e_s^H & 0 \end{bmatrix}.$$

(3) then follows by considering the system

$$\bar{B}\begin{bmatrix} \bar{y} \\ \mu \end{bmatrix} = \begin{bmatrix} r+y_s\bar{y} \\ 0 \end{bmatrix}.$$

Introduce the following shorthand notation

$$Z_\lambda = Z - \lambda E,$$

$$z_{\lambda s} = \text{the } s(\text{th}) \text{ column of } Z - \lambda E.$$

We can express the matrix B in terms of these new quantities,

$$B = A_\lambda - (x+a_{\lambda s})e_s^H.$$

Let $c = -x - a_{\lambda s}$ and $g = r + y_s\bar{y}$.

With these expressions (3) becomes

$$By = (A_\lambda+ce_s^H)y \tag{5}$$
$$= g.$$

Unfortunately the matrix B formed above is singular if and only if λ is a multiple eigenvalue of A, thus we can only improve single eigenvalues; see Appendix.

The Schur factorisation of A is

$$A = QTQ^H$$

We will show how this can be used to solve the system (4) in $O(n^2)$.

(4) becomes

$$By = (QT_\lambda Q^H+ce_s^H)y$$
$$= Q(T_\lambda Q^H+Q^H ce_s^H)y \tag{6}$$
$$= Q(T_\lambda+Q^H ce_s^H Q) \ Q^H y$$

If $d = Q^H c$, and $f^H = e_s^H Q$ then (5) can be written as

$(T_\lambda + df^H)Q^H y = Q^H g.$

We now construct an orthogonal matrix $Q_1 = P_2 \ldots P_n$, where each P_i acts upon d eliminating each d_i.

As we are eliminating the elements of d, we apply the same transformations on T_λ.

Consider the 4 by 4 case

$$
\begin{bmatrix} d_1 \\ d_2 \\ d_3 \\ d_4 \end{bmatrix} \quad \text{and} \quad
\begin{bmatrix}
t_{1,1}-\lambda & t_{1,2} & t_{1,3} & t_{1,4} \\
0 & t_{2,2}-\lambda & t_{2,3} & t_{2,4} \\
0 & 0 & t_{3,3}-\lambda & t_{3,4} \\
0 & 0 & 0 & t_{4,4}-\lambda
\end{bmatrix}
$$

After pre-multiplication by P_4, which only acts on d_3 and d_4, the system will look like:

$$
\begin{bmatrix} d_1 \\ d_2 \\ X \\ 0 \end{bmatrix} \quad \text{and} \quad
\begin{bmatrix}
t_{1,1}-\lambda & t_{1,2} & t_{1,3} & t_{1,4} \\
0 & t_{2,2}-\lambda & t_{2,3} & t_{2,4} \\
0 & 0 & X & X \\
0 & 0 & X & X
\end{bmatrix}
$$

i.e. a subdiagonal element will have filled in. The process continues until d has become a multiple of e_1 and $T_\lambda + df^H$ has become upper Hessenberg.

The remaining step is to form a Q_2 which will 'drive off' the subdiagonal elements. Thus, we are left with the tridiagonal system of equations to solve.

$$T'_\lambda \, Q^H \, y = Q_2 \, Q_1 \, Q^H \, g \tag{7}$$

Equation (1) which leads to (7) is nonlinear. The process is repeated with $x+\tilde{y}$ and $\lambda+\mu$ as new approximations. The theorem establishing convergence for this iterative procedure is given in [5]. Since an orthogonal triangularization of A is available, the matrix B can be updated at each stage of the iteration, using the current approximation to the eigenpair. So we treat the $(p+1)$th step as though it was the first step in the iteration with starting values λ^p and μ^p. Consider (4), this can be rewritten as

$$y = B^{-1}r + B^{-1}y_s\tilde{y}. \tag{8}$$

The iteration scheme is

$$y^{p+1} = B^{-1}r + B^{-1} \, y^p_s \, \tilde{y}^p, \tag{9}$$

but as we are treating the $(p+1)$th iteration as the first, the second term above becomes zero, i.e.

$$y^{p+1} = B^{-1}r \tag{10}$$

So (7) can thus be written as

$$T'_\lambda \, Q^H \, y = Q_2 \, Q_1 \, Q^H \, r \tag{11}$$

Equation (11) is solved for each stage of the iteration. So after solution of the above equation, T'_λ must be transformed back to T_λ. The rotations to do this are stored in d and an auxillary vector. Note that gaussian elimination can be used, but as it may be necessary to pivot, the information about the row iterchanges will have to be kept in two more vectors.

The implementation of this algorithm allows the user to perform one cycle of the iteration or iterate until convergence or the maximum number of iterations has been reached. Thus error bounds can be returned if required.

2.2. Description of the Enclosure Algorithm

Let C, VC, MC be the sets of complex numbers, vectors of complex numbers and matrices of complex numbers respectively. We also have to consider the following roundings $\{\square, \nabla, \triangle, \Diamond\}$ These are a monotonic antisymmetric rounding, downwardly and upwardly directed roundings and rounding to the nearest enclosing interval respectively.

As the purpose of this paper is to present information not previously covered, we merely state the following enclosure theorem and give a reference to its proof.

Theorem

Let $A \in M_n C, R \in M_{n+1} C, \bar{x} \in V_n C$ and $\bar{\lambda} \in C$. For $Y \in IV_n C$ and $M \in IC$ the function $F_{\Diamond} : IV_{n+1} C \to IV_{n+1} C$ is defined by

$$F_{\Diamond}\begin{bmatrix} Y \\ M \end{bmatrix} := \Diamond R \Diamond\Diamond \begin{bmatrix} A \cdot \bar{x} - \bar{\lambda} \bar{x} \\ 0 \end{bmatrix} \Diamond\Diamond E_{n+1} - R \cdot \begin{bmatrix} A - \bar{\lambda} E & -\bar{x} - Y \\ e'_s & 0 \end{bmatrix} \Diamond \begin{bmatrix} Y \\ M \end{bmatrix}$$

If for some $U \in IV_{n+1} C$

$$F_{\Diamond} \overset{\circ}{\subset} U \text{ with } U = \begin{bmatrix} X \\ \Lambda \end{bmatrix},$$

then there exists at least one eigenvalue $\hat{\lambda} \in \bar{\lambda} \Diamond \Lambda$ and at least one eigenvector $\hat{x} \in x \Diamond X$ with $A \cdot \hat{x} = \hat{\lambda} \cdot \hat{x}$. The eigenvalue $\hat{\lambda}$ possesses the multiplicity one. Proof [2]. As usual the matrix R is chosen as the inverse of

$$\begin{bmatrix} A - \bar{\lambda} \cdot E & -\bar{x} \\ e'_s & 0 \end{bmatrix}$$

We have seen that this system is equivalent to B in the previous section, see equation (3), so therefore as before we reduce the problem of order $(n+1)$ to one of order n. We also saw an economical method for computing its effective inverse. This method was modified to produce the actual inverse economically for use in this step. Starting with the interval

$$Y^0 = -R \Diamond\Diamond ((A - \bar{\lambda}) \cdot \bar{x}) \text{ the intervals } Y^l, l{\neq}1, \text{ are found from}$$

$$Y^1 = Y^0 \Diamond\Diamond (E - R \cdot (X^l)) \Diamond Y^l.$$

until an l has been found such that $Y^{l+1} \overset{\circ}{\subset} Y^l$. The intervals X^l are constructed from Y as

$$x'_i := \begin{cases} y^l & i{\neq}k \\ 0 & i=k \end{cases}$$

In order to improve the speed of convergence, each interval is expanded by a small amount, hopefully including nearby fixed-points. The following technique was used.

Define SIGMA : REAL := NEXT_AFTER(1.0,1.5); the function NEXT_AFTER returns the next machine number after the first argument in the direction of the second argument.

An interval A is expanded by A := A * SIGMA;

2.3. The Complete Algorithm

A. Find eigenpair, if the system is poorly conditioned then we will find an approximate eigenpair.

The following steps are performed in order to obtain a good 'approximate' eigensolution and the Schur factorization.

1. Reduction of A to Hessenberg form H via unitary transformations and returning Q the matrix of accumulated transformations.

2. Computation of the eigenvalues and the Schur factorization by a modified QR algorithm delivering the upper-triangular matrix T and updating matrix Q with the unitary transformations used in the reduction.

3. Computation of the selected eigenvector using inverse iteration.

4. Back-transformation of the eigenvector of T using Q.

5. Improving the eigenpair by iterative refinement. As the version of iterative refinement used can accept a poor approximation to the eigenvector, steps 3 and 4 can be omitted.

6. Obtain R economically.

Note: if the nature of the problem permits it, then balance A before reduction to Hessenberg form, remembering that it is not possible to recover the eigenvectors of the original matrix A as balancing is a similarity transformation and we are using orthogonal transformations in our reduction.

B. Verify the eigensolution.

-- Try to catch fixed point by successively expanding interval.

p:=0

loop

$Y^p := Y^p * SIGMA;$

$Y^{p+1} = Y^0 \lozenge\lozenge (E - R \cdot B(X^p)) \lozenge Y^p$

Enclose := $Y^{p+1} \overset{\circ}{\subset} Y^p$

p:=p+1;

exit when Enclose or p > Max_Iterations or Interval_too_wide;

end loop;

The matrix $B(X^p) := \left[(A - \bar{\lambda}E)_1 \; \dots \; (A - \bar{\lambda}E)_{k-1} \; -\bar{x} - X^l \; (A - \bar{\lambda}E)_{k+1} \; \dots \; (A - \bar{\lambda}E)_n \right]$

and $\bar{x}_k := \dfrac{\max}{1 \le i \le n} \bar{x}_i.$

3. Implementation

Ada was chosen as the implementation language because it allows both generics and overloading of operators. Generics allow the same library unit to supply the user with programs of various precision or even with a program which operates on a different variable type for which the algorithm still has a sensible meaning. By using overloaded operators the algorithm in the source file can correspond more closely with the text book version. In addition, Ada programs are much more portable than those of other languages. This is important when one is trying to reach a wide audience who possess a variety of machine types.

The above verification algorithm is valid with respect to real and complex numbers. The generic facility of Ada allows a procedure which deals with these two cases to be delivered.

3.1. Specification

```
--  ==========================================================================
--
--  ABSTRACT:
--
--  PERFORMS VERIFICATION OF EIGENSOLUTIONS.
--  THIS PROCEDURE IS GENERIC WRT REAL AND COMPLEX TYPES.
--  NOTE: IT WILL NOT VERIFY COMPLEX EIGENSOLNS OF REAL MATRICES.
--
--  ==========================================================================
generic

    type FLOAT_TYPE is digits <>;
    --
    type SCALAR_TYPE is private;
    type SCALAR_VECTOR_TYPE is array (INTEGER range <>) of SCALAR_TYPE;
    type SCALAR_MATRIX_TYPE is array (INTEGER range <>, INTEGER range <>)
                                                     of SCALAR_TYPE;
    --
    ONE: SCALAR_TYPE;
    --
    with function ABSOLUTE(X: SCALAR_TYPE) return FLOAT_TYPE is <>;
    --
```

```
with function NEGATE(X:SCALAR_TYPE) return SCALAR_TYPE is <>;
with function ADD(X,Y:SCALAR_TYPE) return SCALAR_TYPE is <>;
with function SUBTRACT(X,Y:SCALAR_TYPE) return SCALAR_TYPE is <>;
with function MULTIPLY(X,Y:SCALAR_TYPE) return SCALAR_TYPE is <>;
with function DIVIDE(X,Y:SCALAR_TYPE) return SCALAR_TYPE is <>;
--
type SCALAR_INTERVAL_TYPE is private;
type SCALAR_INTERVAL_VECTOR_TYPE is array(INTEGER range <>) of
                                    SCALAR_INTERVAL_TYPE;
--
with function COMPOSE_INTERVAL(X,Y:SCALAR_TYPE)
        return SCALAR_INTERVAL_TYPE is <>;
with function "-"(X:SCALAR_INTERVAL_TYPE)
        return SCALAR_INTERVAL_TYPE is <>;
with function "+"(X,Y:SCALAR_INTERVAL_TYPE)
        return SCALAR_INTERVAL_TYPE is <>;
with function "-"(X,Y:SCALAR_INTERVAL_TYPE)
        return SCALAR_INTERVAL_TYPE is <>;
with function "*"(X,Y:SCALAR_INTERVAL_TYPE)
        return SCALAR_INTERVAL_TYPE is <>;
with function "<="(X,Y:SCALAR_INTERVAL_VECTOR_TYPE)
        return BOOLEAN is <>;
with function COMPOSE_INTERVAL(X,Y:SCALAR_VECTOR_TYPE)
        return SCALAR_INTERVAL_VECTOR_TYPE is <>;
--
type SCALAR_DOT_PRECISION_TYPE is limited private;
--
with procedure CLEAR(C:in out SCALAR_DOT_PRECISION_TYPE) is <>;
with procedure ACCUMULATE(C:in out SCALAR_DOT_PRECISION_TYPE;
                        X:SCALAR_TYPE) is <>;
with procedure DOT_ADD(C:in out SCALAR_DOT_PRECISION_TYPE;
                        X,Y:SCALAR_VECTOR_TYPE) is <>;
with function ROUND_DOWN(C:SCALAR_DOT_PRECISION_TYPE)
        return SCALAR_TYPE is <>;
with function ROUND_UP(C:SCALAR_DOT_PRECISION_TYPE)
        return SCALAR_TYPE is <>;
--
type SCALAR_INTERVAL_DOT_PRECISION_TYPE is limited private;
--
with procedure CLEAR(C:in out SCALAR_INTERVAL_DOT_PRECISION_TYPE)
                                                    is <>;
with procedure ACCUMULATE
                (C:in out SCALAR_INTERVAL_DOT_PRECISION_TYPE;
                 X:SCALAR_INTERVAL_TYPE) is <>;
with procedure DOT_ADD(C:in out SCALAR_INTERVAL_DOT_PRECISION_TYPE;
                        X,Y:SCALAR_INTERVAL_VECTOR_TYPE) is <>;
with function ROUND(C:SCALAR_INTERVAL_DOT_PRECISION_TYPE)
        return SCALAR_INTERVAL_TYPE is <>;

procedure GENERIC_ENCLOSE_EIGENSOLUTION(
                        A          : SCALAR_MATRIX_TYPE;
                        LAMBDA     : SCALAR_TYPE;
                        X          : SCALAR_VECTOR_TYPE;
                        INT_LAMBDA : out SCALAR_INTERVAL_TYPE;
                        INT_X      : out SCALAR_INTERVAL_VECTOR_TYPE;
                        MAXITS     : POSITIVE;
                        FAIL       : out INTEGER );
```

3.2. Usage

An instance which verifies real eigensolutions of real matrices is shown below. The procedure is instantiated in a package to allow visibility of the declarations contained within the subpackages of GENERIC_SCIENTIFIC_COMPUTATION and instances thereof. Without this nesting, the type declarations would have to be accessed using the dot notation. This is felt to be to painful and a little unreadable.

```
with NAG_A01AA_REAL;use NAG_A01AA_REAL;
--
-- IMPORT REAL TYPES.
--
with REAL_SCIENTIFIC_COMPUTATION;
use REAL_SCIENTIFIC_COMPUTATION;
--
-- IMPORT AN INSTANTIATION OF GENERIC_SCIENTIFIC_COMPUTATION.
--
with GENERIC_ENCLOSE_EIGENSOLUTION;
package VERIFY_REAL_EIGENSOLUTION is

    use REAL_SCIENTIFIC_COMPUTATION.REAL_ARITHMETIC;
    use REAL_SCIENTIFIC_COMPUTATION.INTERVAL_ARITHMETIC;

    procedure VERIFY_EIGENSOLUTION is new
    GENERIC_ENCLOSE_EIGENSOLUTION
    (REAL,
     REAL,
     REAL_VECTOR,
     REAL_MATRIX,
     ONE=>1.0,
     SCALAR_INTERVAL_TYPE=>INTERVAL,
     SCALAR_INTERVAL_VECTOR_TYPE=>INTERVAL_VECTOR,
     SCALAR_DOT_PRECISION_TYPE=>DOT_PRECISION,
     SCALAR_INTERVAL_DOT_PRECISION_TYPE=>INTERVAL_DOT_PRECISION);

end;
```

4. Appendix

Theorem: The matrix $B = A_\lambda - (x + a_{\lambda s}) \, e_s^H$ is singular if and only if λ is a multiple eigenvalue of A.

Dongarra in [5] attributes this theorem to Wilkinson. It is repeated here, though written in terms of complex arithmetic.

B is singular if λ is a multiple eigenvector of A.

Proof: If λ and x are the exact eigenpair and λ is a multiple eigenvalue of A, then there exists at least one left eigenvector y belonging to λ which is orthogonal to x. Hence

$$y^H (A - \lambda E) = 0 \text{ and } y^H x = 0.$$

Multiply y^H by (4), to give

$$y^H B = y^H (A - \lambda E + ce_s^H) = 0,$$

i.e. y is a null solution of B, so B is singular.

On the other hand, if B is singular then there exists a $z \neq 0$ such that

$$Bz = 0 \text{ or}$$

form $\bar{z} = (z_1 ... z_{s-1}, 0, z_{s+1} ... z_n)$ so we can say $B(\bar{z} + z_s e_s) = 0$.

This implies that

$$B\bar{z} = z_s x \text{ because } Be_s = -x.$$

But $B\bar{z} = (A-\lambda E)\bar{z}$ since $z_s = 0$. Hence

$$(A-\lambda E)\bar{z} = z_s x.$$

If $z_s = 0$, then $\bar{z}$ is an eigenvector of A corresponding to λ. Because $x_s = 1$ and $z_s = 0$, the vectors x and $\bar{z}$ are independent. If z_s. Thus we have

$$(A-\lambda E)^2 \bar{z} = 0.$$

Hence $\bar{z}$ is a principle vector of grade 2. In either case λ has multiplicity of at least 2.

5. Glossary

E	....	The unit matrix
y_s	....	The s(th) component of vector Y
e_s	....	The s(th) column of unit matrix
x^t	....	The transpose of x
x^H	....	The Hermitian transpose of x

6. References

[1] Erl, M., Hodgson, G., Kok, J., Winter, D. and Zoellner, A.
Design and Implementation of Accurate Operators in Ada.
Diamond Deliverable D1-2/4, Doc No. 03/1-2/1/S02.f

[2] Gruñer, K.
Solving The Complex Eigenvalue Problem with Verified High Accuracy.
Diamond Deliverable D3-1/2, Doc No. 03/3-2/1/K1.f

[3] Wilkinson, J.H.
The Algebraic Eigenvalue Problem.
Clarendon Press, Oxford, 1965.

[4] Wilkinson, J.H. and Reinsch, C.
Handbook for Automatic Computation, (Vol. 2, Linear Algebra).
Springer Verlag, Berlin, 1971.

[5] Dongarra, J.J., Moler, C.B. and Wilkinson, J.H.
'Improving the accuracy of computed eigenvalues and eigenvectors'.
SIAM J. Numer. Anal., 20, 1, 1983.

[6] Rump, S.M.
'Solution of linear and nonlinear algebraic problems with sharp, guaranteed bounds'.
Computing Suppl., 5, pp. 147-168, 1984.

Enclosing all Eigenvalues of Symmetric Matrices

DIAMOND Deliverable D3-1 part 3
Rudolf Lohner
Universität Karlsruhe

Abstract: In this paper a method is presented which computes highly accurate enclosures for the eigenvalues of a symmetric real matrix $A = A^T$. This method rests on the following basic steps: (i) computation of eigenvalue and eigenvector approximations using the classical JACOBI method, (ii) computing an enclosure of the transformed matrix $A_1 = T^{-1}AT$ where the columns of T are the eigenvector approximations and (iii) computing eigenvalue enclosures by use of GERSCHGORIN's theorem for A_1. To gain more accuracy in the results this basic method is supplemented by suitable additional steps and tools such as reorthonormalization, a slightly modified JACOBI method and an exact scalar product on the computer.

1 Introduction

Considering the eigenvalue problem for a real, symmetric n×n matrix A such that

$$A x = \lambda x,$$

it is well known that there exist n real eigenvalues (including multiplicities) and a set of n orthogonal eigenvectors, i.e. there exists a real orthogonal matrix T which transforms A into a diagonal matrix D:

$$D = T^{-1} A T, \quad \text{where } T^{-1} = T^H.$$

The columns of T are the eigenvectors of A and the diagonal of D consists of the eigenvalues of A.

Many iterative methods are known which can be used to compute approximations of the eigenvalues and eigenvectors of A. (The JACOBI-, QR- and LR-methods are only the most important ones, see [2],[12],[13],[14],[15]). Generally, these methods compute a series of similarity transformations starting with the matrix A and thereby overwriting A by the transformed matrices. However, since the transformed matrices are perturbed by roundoff errors, the original problem is more and more perturbed and eventually the computed eigenvalues and eigenvectors can differ considerably from the exact quantities. Unless we feed the original matrix A back into the computation, there is no hope that we can get around this accumulation of errors.

If we want to control the errors during computation in order to get enclosures of the eigenvalues and eigenvectors, we can use for example a residual iteration as is done by RUMP [10]. There the residual of an eigenvalue/eigenvector approximation (using the original matrix A) is used in connection with interval arithmetic to control and eventually enclose the error. This method, which is essentially NEWTONS method for nonlinear systems, can be used for arbitrary matrices and it can produce enclosures of very high accuracy. However, it only works if the eigenvalue under consideration is simple. Also, for clusters of (simple) eigenvalues, this method can break down because of ill conditioning of the problem. This approach is also used in [3], [6] and [9].

In this paper, we will demonstrate that enclosures of high accuracy can also be obtained in the case of symmetric matrices A by use of a classical iterative method (the JACOBI-method) if it is suitably supplemented by interval arithmetic and feedback of the matrix A into the computation. A generalization of the method to complex hermitian matrices does not present any difficulties and is straightforward. Similar enclosure methods can also be developped on the basis of QR-methods or others.

The advantages of this approach are that it can be used to enclose multiple eigenvalues, that clusters of eigenvalues can be treated much better than with the residual iteration method and that the enclosures can be improved as far as desired (this is limited only by underflow). Although the eigenvalue and eigenvector approximations have to be stored with multiple precision, it is not necessary to do all computations with multiple precision arithmetic. The most costly part (the JACOBI iteration) can in fact *always* be done in single precision.

The basic concepts which are used in the method are:
- the classical JACOBI-method and a modification thereof,
- the "staggered correction" format of storing real numbers in a computer,
- a special way to compute similarity transformations,
- the GRAM-SCHMIDT orthogonalization method (or any other) and
- the GERSCHGORIN circle theorem.

By a suitable combination of these concepts, we will be able to compute almost arbitrary narrow bounds for the eigenvalues of A. This will be shown in the following chapters. In Chapter 2, we outline the basic method which gives approximations of the eigenvectors and enclosures of the eigenvalues. In Chapter 3, we show how the eigenvector approximations can be improved and the JACOBI iteration can be continued even if it stopped with a diagonal matrix in the basic method. Chapter 4 then shows how the improved eigenvector approximations can be used to compute improved eigenvalue enclosures. In Chapter 5, we mention shortly how also eigenvector enclosures can be computed by standard estimates due to WILKINSON [14]. Finally, in Chapter 6, we demonstrate the methods with results which were computed with a PASCAL-SC program.

2 Simple Method for Computing Enclosures of Eigenvalues

In this chapter, we start with a simple method for the computation of eigenvector approximations and eigenvalue enclosures. The full method of the following chapters will then be essentially of the same structure, only with more refinements.

This basic method consists of three steps. First we use the JACOBI method to compute an almost orthogonal transformation matrix T such that

$$A_1 := T^{-1} A T \qquad (1)$$

is essentially a diagonally dominant matrix. This transformation matrix T is the rounded product of all GIVENS matrices which occurred during the JACOBI process (it is not very important which version of the JACOBI method is used, in the PASCAL-SC program we used the cyclic order of annihilations). Thus T contains the eigenvector approximations in its columns (and it is a matrix consisting of machine numbers).

Since A_1 and A are similar matrices, they both have the same eigenvalues and since A_1 is much more diagonally dominant than A is, the use of GERSCH-GORIN's theorem for A_1 yields, in general, good enclosures for the eigenvalues of A. However, the matrix A_1 is not known exactly because of the roundoff errors which have occurred during the JACOBI process. Therefore, in the second step of the basic method, we compute the transformation (1) explicitly using the matrix T. The resulting matrix A_1 is not representable with machine numbers in general, therefore we compute the transformation in interval arithmetic to get an enclosure $[A_1]$ of A_1.

This can essentially be done in two ways. The first way is to rewrite (1) as as linear system with coefficient matrix T:

$$T A_1 = A T . \qquad (2)$$

Now an equation solver for linear systems can be used (e.g. [5], [8], [10]) to compute enclosures for the columns of A_1. The fact that (2) contains scalar products in each component of the right hand side requires only a very slight and obvious modification in these solvers, since the residual $b - A\tilde{x}$ of an approximate solution $\tilde{x}$ for Ax=b which occurs in these solvers is computed by an exact scalar product anyway. So, only b has to be replaced by another scalar product. The application of a linear system solver to each column of (2) then gives an enclosure $[A_1]$ of the transformed matrix A_1.

The second way to compute an enlosure $[A_1]$ is to compute an enclosure $[T^{-1}]$ of T^{-1} and then compute

$$[A_1] := [T^{-1}](AT) \tag{3}$$

in interval arithmetic. This is not an inefficient way. Since T is almost orthogonal, and this fact can be used to get $[T^{-1}]$ very quickly.

For an almost orthogonal matrix T, the transposed matrix $R := T^H$ is a good approximation of T^{-1}. In fact, it turns out that in our case the residual matrix $I - RT$ is almost always close to the machine accuracy eps. Thus, the error estimation of the NEUMANN series should give good bounds for the error $\| T - R \|$. This standard error estimation is given by

$$\| T^{-1} - R \| \leq \| R \| \frac{\| I - R T \|}{1 - \| I - R T \|} =: \delta \tag{4}$$

if $\| I - R T \| < 1$. As already mentioned, $\| I - R T \|$ is usually of order eps in our case and $\| R \|$ is of order 1 since R is almost orthogonal. Thus, (4) is a very good estimation of the error.

An enclosure $[T^{-1}]$ of T^{-1} now can be obtained by adding the interval $[-\delta, \delta]$ to each component of $R = T^H$:

$$T^{-1} \in [T^{-1}] := T^H + [-\delta, \delta] \begin{pmatrix} 1 & \cdots & 1 \\ \vdots & & \vdots \\ 1 & \cdots & 1 \end{pmatrix} . \tag{5}$$

Now, $[A_1]$ can be computed by (3).

The third step of the basic method is the application of GERSCHGORIN's theorem to the interval matrix $[A_1]$. Since $[A_1]$ contains A_1, we get enclosures of all eigenvalues of A_1 and thus, also of A. The extension of GERSCHGORIN's theorem to interval matrices is very obvious and the eigenvalue enclosure resulting from the i-th row is

$$\lambda_i \in \left[\underline{a}_{ii} - \sum_{\substack{j=1 \\ i \neq j}}^{n} |[a_{i,j}]| , \ \bar{a}_{ii} + \sum_{\substack{j=1 \\ i \neq j}}^{n} |[a_{i,j}]| \right] . \tag{6}$$

Now we have enclosures for all eigenvalues of A, but in general, these enclosures will not be of maximum accuracy and in the case of clusters, the eigenvalues are not separated, in general, by the enclosures. The extension of the method which will be discussed in the following chapters removes these disadvantages almost completely.

3 Computing Eigenvector Approximations with High Accuracy

The key observation for the improvement of the basic method is the simple fact that the diagonal dominance of A_1 and hence the resulting enclosures will be better if T contains better approximations of the eigenvectors. Therefore, the goal in this chapter will be to improve the eigenvector approximations. One way to do this, certainly, is to do all the computations in higher precision arithmetic. This, however, also requires the JACOBI method to be done in higher precision which is quite costly. Instead, we use the so called "staggered correction" format of storing real numbers in the computer which has the advantage that the only operations which occur are single precision operations and/or exact scalar products ([1], [11]). In staggered correction, a real number a is approximated as a sum of REALs whose summands are stored individually, $a \approx a_1 + \dots + a_m$, where $a_1, \dots a_m$ are REALs and the sum is evaluated using the long accumulator.

Thus in the simplest case, we store the matrix T as sum of two REAL matrices T_1 and T_2 to get eigenvector approximations of double length. To increase the accuracy of the eigenvector approximations we proceed as follows.

First we compute single length eigenvector approximations as in the basic method using the ordinary JACOBI method in single precision. This gives us a REAL matrix $\tilde{T}$ of single length eigenvector approximations and a diagonal matrix $D := \mathrm{diag}(\tilde{\lambda}_1, \dots, \tilde{\lambda}_n)$ where $\tilde{\lambda}_1, \dots, \tilde{\lambda}_n$ are the eigenvalue approximations from the JACOBI method. We call this the "first JACOBI method".

Now $\tilde{T}$ is not exactly orthogonal and therefore $\tilde{A}_1 := \tilde{T}^{-1} A \tilde{T}$ is no longer symmetric. The elements of the matrix $I - \tilde{T}^H \tilde{T}$ are of order eps and the off-diagonal elements of $\tilde{A}_1$ will be of order $\|A\|$ eps in general, but their values will differ considerably from symmetry. So it does not make sense to continue the JACOBI method using $\tilde{A}_1$. Instead we orthogonalize the columns of $\tilde{T}$ and store the resulting matrix T in staggered format $T = T_1 + T_2$ such that T is orthogonal in double length. Now the elements of the matrix $I - T^H T = I - T_1^H T_1 - T_1^H T_2 - T_2^H T_1 - T_2^H T_2$ will be of order eps^2 and the off diagonal elements of the resulting transformed matrix $A_1 := T^{-1} A T$ will differ from symmetry only in the digits at the end of the first mantissa.

Therefore, we store the off diagonal elements of A_1 as single precision REALs and the diagonal elements (which are the eigenvalue approximations) in staggered format as sum of two REALs:

$$A_1 = T^{-1} A T = D + \tilde{A}_1 \tag{7}$$

i.e. D is a diagonal and $\tilde{A}_1$ a full matrix.

Remark: The orthogonalization of $\tilde{T}$ does not present any numerical difficulties or instabilities since $\tilde{T}$ is already almost orthogonal. Thus there is no danger in using the unmodified GRAM-SCHMIDT procedure. Since the resulting matrix T has to be stored in double length ($T_1 + T_2$), all scalar products should be computed using the exact scalar product and some arithmetic operations during this orthogonalization have to be done in double precision or by using the long accumulator.

The matrix $\tilde{A}_1$ now can be computed by

$$\tilde{A}_1 = T^{-1}(AT - TD) = T^{-1}(AT_1 + AT_2 - T_1D - T_2D) \tag{8}$$

or, since for the moment we only need an approximation of it, by

$$\tilde{A}_1 \approx \tilde{\tilde{A}}_1 = T^H(AT - TD) = T^H(AT_1 + AT_2 - T_1D - T_2D) \tag{9}$$

where we can take full advantage of the exact scalar product for the expression in parentheses.

So, feeding back the original matrix A by (7) retrieves the information which has been lost during the first JACOBI method in the sense that the transformed matrix A_1 is again symmetric now and thus the JACOBI method can be continued to improve the eigenvector approximations. We call this continuation the "second JACOBI method".

Continuing the JACOBI method with A_1 resp. its approximation $D + \tilde{\tilde{A}}_1$ now will produce only small angles φ in the GIVENS rotations since $D + \tilde{A}_1$ is almost diagonal already. Therefore we gain more accuracy if we do not compute the rotation matrices Ω_{jk} directly (which are close to the identity matrix) but instead we compute their deviation $W_{jk} = \Omega_{jk} - I$ from identity. So we have:

$$\Omega_{jk} = \begin{pmatrix} 1 & & & & & & 0 \\ & \ddots & & & & & \\ & & 1 & & & & \\ & & & c & & -s & \\ & & & & 1 & & \\ & & & \vdots & & \vdots & \\ & & & 1 & & & \\ & & & s & & c & \\ & & & & & & 1 \\ 0 & & & & & & \ddots \\ & & & & & & & 1 \end{pmatrix}, \qquad W_{jk} = \begin{pmatrix} 0 & & & & & & 0 \\ & \ddots & & & & & \\ & & 0 & & & & \\ & & & c-1 & & -s & \\ & & & & 0 & & \\ & & & \vdots & & \vdots & \\ & & & 0 & & & \\ & & & s & & c-1 & \\ & & & & & & 0 \\ 0 & & & & & & \ddots \\ & & & & & & & 0 \end{pmatrix} \tag{10}$$

$c = \cos\varphi, \quad s = \sin\varphi$.

Since for each rotation matrix we compute its deviation from identity, we should also compute their product Ω in this way, $\Omega = I + W$, since it also will differ only little from the identity matrix. Thus, if $\Omega = I + W$ is the transformation matrix at some stage and Ω_{jk} is the next rotation matrix, we compute $\Omega' = I + W' := \Omega \Omega_{jk}$ as follows:

$$\Omega' = \Omega \Omega_{jk} = (I + W)(I + W_{jk}) = I + W + W_{jk} + WW_{jk} , \tag{11}$$
$$\text{i.e.} \qquad W' = W + W_{jk} + WW_{jk} .$$

This second JACOBI method can still be done essentially in single precision: for the off diagonal elements, which are stored in single precision only, the formulas are identically the same as in the case of a single precision matrix A. If $A^i = D + \tilde{\tilde{A}}^i$ and $A^{i+1} = D + \tilde{\tilde{A}}^{i+1}$ are two successive iterates then we have (writing a_{rs} for the elements of A^i and a'_{rs} for the elements of A^{i+1}):

$$\left.\begin{array}{ll} \left.\begin{array}{l} a'_{rj} = a'_{jr} = a_{rj} + s(a_{rk} - \tau a_{rj}) \\ a'_{rk} = a'_{kr} = a_{rk} - s(a_{rj} + \tau a_{rk}) \end{array}\right\} & \begin{array}{l} r = 1, ... , n \\ r \ne j, k \end{array} \\[2em] a'_{jk} = a'_{kj} = 0 , & \tau = \dfrac{s}{1 + c} \end{array}\right\} \tag{11}$$

and all elements in (11) are the entries of the corresponding matrix $\tilde{\tilde{A}}$.

For the diagonal elements we have because of $a_{rr} = d_{rr} + \tilde{\tilde{a}}_{rr}$ and $a'_{rr} = d_{rr} + \tilde{\tilde{a}}'_{rr}$:

$$\left\{\begin{array}{l} \left.\begin{array}{l} a'_{jj} = a_{jj} + t\, a_{jk} \\ a'_{kk} = a_{kk} - t\, a_{jk} \end{array}\right\} \implies \left\{\begin{array}{l} \tilde{\tilde{a}}'_{jj} = \tilde{\tilde{a}}_{jj} + t\, \tilde{\tilde{a}}_{jk} \\ \tilde{\tilde{a}}'_{kk} = \tilde{\tilde{a}}_{kk} - t\, \tilde{\tilde{a}}_{jk} \end{array}\right. \\[2em] t = \tan\varphi . \end{array}\right. \tag{12}$$

Since $\tilde{\tilde{A}}$ is a single precision matrix, we see that this JACOBI method involves essentially only single precision numbers and therefore it is not more expensive than the basic method. The only place where the cost is slightly higher is the computation of t for each GIVENS rotation, which uses the diagonal elements of A_1 so that the difference $a_{jj} - a_{kk}$ occurring in this computation (see [12]) has to be replaced by $d_{jj} + \tilde{\tilde{a}}_{jj} - d_{kk} - \tilde{\tilde{a}}_{kk}$. This expression should be computed using the long accumulator (which is useful especially if multiple eigenvalues or clusters of eigenvalues exist).

When this second JACOBI method is finished we will have improved eigenvalue approximations in the diagonal of A_1 resp. $D + \tilde{\tilde{A}}_1$ which are of double length now. The resulting transformation matrix $\Omega = I + W$ of this second JACOBI method has to be multiplied with the corresponding matrix T from the first JACOBI method (which has been orthogonalized to double length already : $T = T_1 + T_2$) to get the final transformation matrix $T_f = T\Omega$. In order to obtain T_f in double length we do this computation as follows

$$T\Omega = (T_1 + T_2)(I + W) = T_1 + T_2 + T_1W + T_2W \tag{13}$$

where the right hand side is evaluated using the long accumulator. The result then is stored in staggered format in two REAL matrices T_{1f} and T_{2f} which represent the final transformation matrix T_f:

$$T_f := T_{1f} + T_{2f} \approx T_1 + T_2 + T_1W + T_2W. \tag{14}$$

The columns of this matrix T_f will now be eigenvector approximations of double length and in most cases also of double accuracy. With these approximations, it is then possible to get much tighter enclosures for the eigenvalues which we will discuss in the next chapter.

Remark: In practical computations it turnes out that (at least for ill conditioned matrices) the resulting enclosures will be much better if we do a reorthonormalization of the single length transformation matrix during the first JACOBI method already. In our implementation we therefore split the first JACOBI method into two parts: first we stop with a very crude stopping criterion (say, relative or absolute error of 10^{-3}), then we reorthonormalize the transformation matrix (in single length), and do a transformation according to (1) where the diagonal of the transformed matrix is stored in double length already. Then we continue with the first JACOBI method still computing the transformation matrix T in single length only. When the next stopping criterion is met, we proceed as described above.

4 Computing Eigenvalue Enclosures with High Accuracy

The computation of the eigenvalue enclosures does not differ in principle from what we have done in the basic method of Chapter 1. We have only to take into account that our transformation matrix now is of double length.

We have to compute the transformation

$$A_1 = D + \tilde{A}_1 = T_f^{-1} A T_f. \tag{15}$$

Since D and $T_f = T_{1f} + T_{2f}$ are known (machine) matrices, we can do this by computing $\tilde{A}_1$:

$$\tilde{A}_1 = T_f^{-1}(A T_f - T_f D) = T_f^{-1}(AT_{1f} + AT_{2f} - T_{1f}D - T_{2f}D). \tag{16}$$

However, $\tilde{A}_1$ is not representable with machine numbers in general, and therefore, we have to compute an enclosure $[\tilde{A}_1]$. As in Chapter 1, there are again two possibilities of doing this. First, we can rewrite (16) as a linear system and solve it with a verifying linear system solver:

$$T_f \tilde{A}_1 = A T_f - T_f D = A T_{1f} + A T_{2f} - T_{1f} D - T_{2f} D \tag{17}$$

However, here we have a coefficient matrix of double length which would either require a larger modification of the linear system solver as the modification for the right hand side only (see Chapter 1) or we would have to enclose T_f into an interval matrix of single length and call a linear system solver for interval coefficients. Both ways are not very satisfactory, the first because of the larger overhead, the latter because of possibly less accurate results.

Therefore, we use the second approach which we also favoured in Chapter 1. We compute an enclosure $[T_f^{-1}]$ of the inverse T_f^{-1} using the approximation $R := T_f^H = T_{1f}^H + T_{2f}^H \approx T_f^{-1}$ and the error estimation of the NEUMANN series. Since here T_f is orthogonal to approximately two mantissa lengths, we can expect that, in general, we have $\| I - R T_f \| \approx eps^2$, so that the NEUMANN error estimation should be extremely good. The enclosure $[T^{-1}]$ is as in (5):

$$T_f^{-1} \in [T_f^{-1}] := T_{1f}^H + T_{2f}^H + [-\delta,\delta]\begin{pmatrix} 1 & \cdots & 1 \\ \vdots & & \vdots \\ 1 & \cdots & 1 \end{pmatrix} \tag{18}$$

with δ as in (4) but with the new matrix R. As we see below, there is no need to compute $[T_f^{-1}]$ with higher than single precision, so we store this enclosure as an ordinary interval matrix.

As in (3), we now compute an enclosure $[\tilde{A}_1]$ by

$$[\tilde{A}_1] := [T_f^{-1}] (A T_f - T_f D) = [T_f^{-1}] (A T_{1f} + A T_{2f} - T_{1f} D - T_{2f} D). \tag{19}$$

Here, the expression in parenthesis must be evaluated with an exact scalar product in each component, since there will be a substantial amount of cancellation. The resulting matrix will be an interval matrix of maximum accuracy and the final multiplication with the (orthogonal) matrix $[T_f^{-1}]$ will not introduce any significant overestimation even if $[T_f^{-1}]$ is only of single length (as mentioned above).

With $[A_1] := D + [\tilde{A}_1]$, we now have an enclosure of the transformed matrix A_1 from (15) whose diagonal elements are double length approximations of the eigenvalues and whose off diagonal elements are approximately eps^2 times smaller than the diagonal elements if the eigenvalues are not ill conditioned. The relation of diagonal to off diagonal elements will become worse as the condition of the eigenvalues becomes worse, i.e. as the ratio between the largest and the smallest eigenvalues (in modulus) increases.

As at the end of the basic method in Chapter 1, we finally compute the eigenvalue enclosures by use of the GERSCHGORIN theorem applied to the interval matrix $[A_1]$. The eigenvalue enclosures are then of the form

$$\lambda_i \in \left[d_{ii} + \underline{\tilde{a}}_{ii} - \sum_{\substack{j=1 \\ i \neq j}}^{n} |[\tilde{a}_{i,j}]| \ , \ d_{ii} + \overline{\tilde{a}}_{ii} + \sum_{\substack{j=1 \\ i \neq j}}^{n} |[\tilde{a}_{i,j}]| \ \right] \tag{20}$$

from which it is obvious that they will be of double length accuracy in the well conditioned case and that accuracy will be less than single length if the ratio between off diagonal and diagonal elements of A_1 becomes greater than eps. This will be the case if the ratio of the largest and the smallest eigenvalue increases above 1/eps (see the examples in Chapter 6).

5 Computing Eigenvector Enclosures

Up to now we have only considered the problem of enclosing the eigenvalues. The reason for this is that the method we have presented in chapters 3 and 4 is not standard and therefore we consider it as the main object of this paper. Once we have good enclosures for the eigenvalues we can proceed with standard estimations to obtain also enclosures for the eigenvectors. For a posteriori estimations of the error of approximate eigenvectors see WILKINSON, [14], where for eigenvectors of simple eigenvalues we can find the estimation

$$\| x - \tilde{x} \|_2^2 \leq \frac{\varepsilon^2}{a^2} + \left[1 - \left(1 - \frac{\varepsilon^2}{a^2} \right)^{1/2} \right]^2 \tag{21}$$

for an eigenvector x and a corresponding approximation $\tilde{x}$. Here it is assumed that $\|x\|_2 = \|\tilde{x}\|_2 = 1$ and $\varepsilon := \| A\tilde{x} - \tilde{\lambda}\tilde{x} \|_2$. With a we denote the largest distance of $\tilde{\lambda}$ to the next eigenvalue, i.e. λ and its approximation $\tilde{\lambda}$ are contained in $[\tilde{\lambda}-a, \tilde{\lambda}+a]$ and all the other eigenvalues are outside of this interval (and a should be chosen as large as possible, of course).

For multiple eigenvalues similar estimations hold, however, in this case we cannot decide numerically whether or not there is a multiple eigenvalue or if we merely cannot separate eigenvalues lying close together. Thus, we can also not decide as to what dimensions the corresponding eigenspaces actually have. Therefore, the best we can do is to enclose a basis of the subspace which is spanned by all eigenvectors corresponding to all eigenvalues in an enclosing interval which is disjoint from all the other eigenvalue enclosures but contains more than one eigenvalue.

How this can be done is shown in [7]. This paper also contains a test which proves the enclosure of a basis of this subspace, i.e. that any set of vectors out of these enclosures form a linearly independent set. The subspace itself must not necessarily be an eigenspace; it will be an eigenspace if and only if the corresponding eigenvalue enclosure contains only one multiple eigenvalue (which, in general, we cannot decide numerically).

6 Numerical Examples

The method described in the previous sections was applied successfully to many matrices from [4] and to others. One advantage of the method is that it *always gives enclosures for all eigenvalues*. It is only the accuracy of the results which differs with the condition of the problem. Experience with numerical tests showes that for well conditioned eigenvalues the relative accuracy of the enclosures is always more than t digits where t is the length of the mantissa of the underlying floating-point arithmetic. If eps is the machine accuracy, λ_{max} is the (absolutely) largest eigenvalue of a matrix, and for any other eigenvalue λ of this matrix we have $|\lambda/\lambda_{max}| <$ eps, then the enclosure for λ will not be of maximum accuracy in general and the accuracy will decrease further as $|\lambda/\lambda_{max}|$ further decreases. For $|\lambda/\lambda_{max}| <$ eps^2, the enclosure for λ will usually contain zero. The absolute accuracy of the results is for most cases roughly $|\lambda_{max}|$eps^2. These observations can be confirmed in the examples of this section.

All examples were computed using an ATARI ST computer and the programming language PASCAL-SC 2 with a 13 digit decimal arithmetic and an exact scalar product. From the observations made above, it follows that for well conditioned eigenvalues we should expect enclosures of maximum accuracy; for $|\lambda/\lambda_{max}| < 10^{-13}$, the enclosure of λ will have an accuracy of less than 13 digits and for $|\lambda/\lambda_{max}| < 10^{-26}$, there will be no significant digit in the enclosure for λ, then in general, this enclosure will contain zero.

Example 1 (multiple eigenvalues)

This simple example demonstrates the fact that our method can enclose multiple eigenvalues (which is not possible with defect correction).

Let

$$
A_n := \begin{pmatrix} a+1 & 1 & \ldots & 1 & 1 \\ 1 & & & & \vdots \\ \vdots & & \ddots & & 1 \\ \vdots & & & & 1 \\ 1 & 1 & \ldots & 1 & a+1 \end{pmatrix} = \begin{pmatrix} 1 & \ldots & 1 \\ & & \\ \vdots & & \vdots \\ & & \\ 1 & \ldots & 1 \end{pmatrix} + a \cdot I
$$

where $a \in R$ arbitrary (such that a+1 can be stored exactly in the computer). The eigenvalues of A_n are $\lambda_1 = \ldots = \lambda_{n-1} = a$, λ_n = n+a. Results were computed for various $n \le 20$. For $|a| \ge 10^{-12}$ the computed enclosures were always of maximum accuracy (13 digits). For a = 0, the enclosure of λ_n was of maximum accuracy and the enclosures of $\lambda_1, \ldots, \lambda_{n-1}$ were, in general, smaller than $[-10^{-24}, 10^{-24}]$.

Example 2 (HILBERT matrix, large ratio of large and small eigenvalues)

We compute the eigenvalues of the n-dimensional HILBERT matrix A_n which is multiplied by a suitable integer m (the lcm of all denominators) to make the matrix entries exactly storable. A_n is defined as

$$A_n = \left(a_{ij} \right) = \left(\frac{1}{i+j-1} \right).$$

In the two cases n = 11 and n = 18 which we will consider, it is multiplied by m = 232 792 560 and m = 144 403 552 893 600.

The following tables contain the eigenvalue enclosures of these two matrices. We can see very clearly the effect on ill conditioned eigenvalues. The accuracy decreases until no correct digit is obtained when the ratio of the largest and the smallest eigenvalue approaches $1/\text{eps}^2 = 10^{26}$ for the arithmetic used in these examples (see λ_1 for n = 18).

Of course, in those cases for which 12 digits are correct, the accuracy still increases internally in the algorithm, it is only the output which is rounded to 13 digit intervals. Thus, for the largest eigenvalues we have in both cases an internal accuracy of approximately 24 - 25 digits !

Table 1, n = 11

eigenvalue	inclusion	number of correct digits
λ_1	7.899160434^{88}_{79} E-07	10
λ_2	1.81742986496^{6}_{4} E-04	12
λ_3	1.92829732741^{9}_{8} E-02	12
λ_4	1.24971854302^{7}_{6} E+00	12
λ_5	5.5212795964^{30}_{29} E+01	12
λ_6	1.755815982827^{8}_{7} E+03	12
λ_7	4.13040377451^{3}_{2} E+04	12
λ_8	7.25017140086^{3}_{2} E+05	12
λ_9	9.38378002941^{8}_{7} E+06	12
λ_{10}	8.43598634381^{9}_{8} E+07	12
λ_{11}	4.13179599056^{7}_{6} E+08	12

99

Table 2, n = 18

eigenvalue	inclusion	number of correct digits
λ_1	[5.7E-12, 1.9E-11]	0
λ_2	4.76^{9}_{2} E-09	3
λ_3	8.837^{88}_{79} E-07	4
λ_4	1.040265^{96}_{70} E-04	7
λ_5	8.7224948^{60}_{51} E-03	8
λ_6	5.5430398847^{7}_{1} E-01	11
λ_7	2.772928246^{30}_{28} E+01	11
λ_8	1.11948098860^{4}_{3} E+03	12
λ_9	3.70925117029^{4}_{3} E+04	12
λ_{10}	$1.0201140010 5^{4}_{3}$ E+06	12
λ_{11}	$2.3451737223 4^{8}_{7}$ E+07	12
λ_{12}	$4.5223426995 2^{1}_{0}$ E+08	12
λ_{13}	$7.3138397012 6^{8}_{7}$ E+09	12
λ_{14}	9.8759016585^{70}_{69} E+10	12
λ_{15}	1.10130177724^{5}_{4} E+12	12
λ_{16}	9.90161430679^{5}_{4} E+12	12
λ_{17}	$6.7126846443 0^{2}_{1}$ E+13	12
λ_{18}	$2.7223185553 4^{8}_{7}$ E+14	12

These two matrices were also treated as test examples in [3] where a binary arithmetic with 53 bits mantissa was used ($\approx$ 16 decimal digits) in a NEWTON iteration. Whereas, according to this higher working precision, the results there were more accurate for n = 11 and for $\lambda_5, \ldots, \lambda_{18}$ if n = 18, this is no longer the case for $\lambda_1, \ldots, \lambda_4$ if n = 18 : not even the sign of these four eigenvalues can be determined from the enclosures obtained there. Although our results are not of maximum accuracy, they still show the positive definiteness of A_{18}.

This example illustrates that ill conditioned eigenvalues do not lead to a breakdown of the method or to a sudden drastic loss of accuracy as is the case for defect correction methods or NEWTON's method. The loss of accuracy is fluent but the method never breaks down. If the accuracy should not be sufficient, then the method can easily be extended to yield higher accuracy by taking more than two mantissas for the data which is stored in staggered correction form in the algorithm.

Example 3 (modified ZIELKE matrix, clusters of eigenvalues)

Consider the following symmetric n×n matrix

$$
A_n = \begin{pmatrix}
1+\varepsilon & 1+\varepsilon & 1+\varepsilon & \cdots & 1+\varepsilon & 1 \\
1+\varepsilon & 1+\varepsilon & & & 1 & 1 \\
1+\varepsilon & & & & & \vdots \\
\vdots & & & & & 1 \\
1+\varepsilon & 1 & & & 1 & 1 \\
1 & 1 & \cdots & 1 & 1 & 1-\varepsilon
\end{pmatrix}
\qquad \text{with} \quad \varepsilon = 10^{-12}
$$

which is obtained by a small modification of a ZIELKE matrix (see [16]). This matrix is exactly storable in the 13-digit decimal arithmetic of PASCAL-SC. It is essential that ε is chosen such that A_n is exactly storable, since there exists an eigenvalue which is very sensitive to changes in A_n (see example 4).

The eigenvalues of A_n are not known explicitly, however, the computed enclosures of the eigenvalues (for $2 \le n \le 20$) display the following behaviour:

$$
\lambda_1 \approx -5 \cdot 10^{-25}, \qquad \lambda_n \approx n, \qquad |\lambda_k| \approx 10^{-12} \text{ for } 2 \le k \le n-1 .
$$

Figure 1 shows the results of the traditional JACOBI method as well as the new enclosure method. For each $n = 3, ..., 20$, this figure contains a horizontal line where the upward ticks ($\perp$) denote the approximations of the eigenvalues by the JACOBI method and the downward ticks ($\top$) denote the true eigenvalues obtained from the enclosures. The values in Figure 1 are plotted on a logarithmic scale with the largest eigenvalue ($\lambda_n \approx n$) and the smallest one ($\lambda_1 \approx -5 \cdot 10^{-25}$) omitted.

As can be seen in this figure, the approximations of the traditional JACOBI method are for the most part correct in the order of magnitude, their values however are more or less random numbers. Moreover, in none of the computed cases there is any indication that there exists an eigenvalue near $-5 \cdot 10^{-25}$. The absolute values of all approximations (except the one for $\lambda_n \approx n$) vary between 10^{-13} and 10^{-11}. (This is the reason why in Figure 1 there is always one upward tick more than the corresponding downward ticks).

The enclosures for $\lambda_2, ..., \lambda_{n-1}$, on the other hand, are all correct to at least 10 digits; λ_n is always of least bit accuracy and $\lambda_1 \approx -5 \cdot 10^{-25}$ is the only eigenvalue which is not obtained with high relative accuracy, the enclosures for this eigenvalue are in general contained in the interval $[-10^{-24}, -10^{-25}]$.

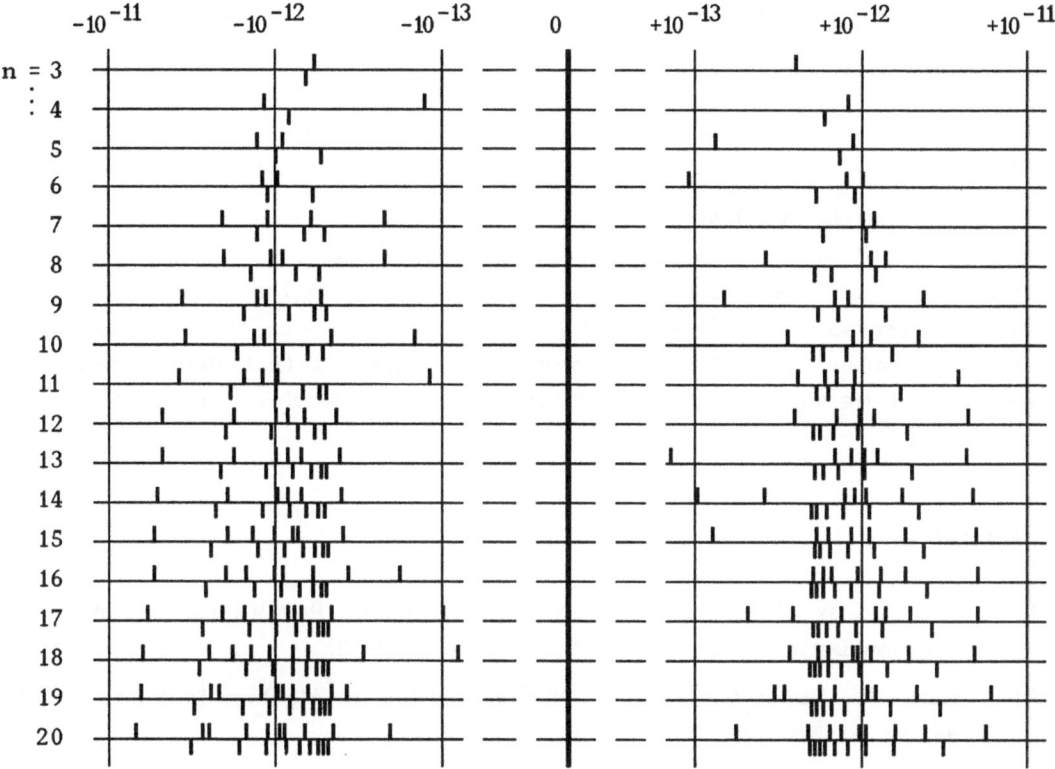

Figure 1. Enclosures ($\top$) and JACOBI approximations ($\bot$) for the eigenvalues of the modified ZIELKE matrix A_n.

Example 4 (modified ZIELKE matrix, very sensitive eigenvalue)

Here we take the matrices from Example 3 but change the element a_{nn} from $1-\varepsilon$ to 1 which is an absolute and relative change of $\varepsilon = 10^{-12}$. So the matrix is now :

$$
A_n = \begin{pmatrix}
1+\varepsilon & 1+\varepsilon & 1+\varepsilon & \dots & 1+\varepsilon & 1 \\
1+\varepsilon & 1+\varepsilon & & & 1 & 1 \\
1+\varepsilon & & & & & \vdots \\
\vdots & & & & & 1 \\
1+\varepsilon & 1 & & & 1 & 1 \\
1 & 1 & \dots & 1 & 1 & 1
\end{pmatrix}
\qquad \text{with} \quad \varepsilon = 10^{-12}.
$$

For $2 \leq n \leq 10$ we computed enclosures for the eigenvalues with the result that

$$|\lambda_k| \approx 10^{-12} \text{ for } k = 1, \ldots, n-1, \quad \text{and} \quad \lambda_n \approx n$$

where λ_n was of maximum accuracy and $\lambda_1, \ldots, \lambda_{n-1}$ were accurate to at least 11 digits.

We see that the smallest eigenvalue $\lambda_1 \approx -5 \text{ E-25}$ from Example 3 has disappeared and changed by a factor of $\approx 10^{13}$! Since the change of a_{nn} in the matrix has a relative and absolute error of only 10^{-12}, this means that in Example 3, the smallest eigenvalue λ_1 is extremely sensitive with respect to a_{nn} and has a condition number of $\approx 10^{25}$! Therefore it is expected that defect correction methods and NEWTON's method will have strong difficulties in treating these matrices.

7 References

[1] Auzinger, W., Stetter, H. J., Accurate Arithmetic Results for Decimal Data and Non-Decimal Computers, Computing 35, 1985.

[2] Bunse, W., Bunse-Gerstner, A., Numerische lineare Algebra. Teubner Studienbücher Mathematik, B.G. Teubner, Stuttgart, 1985.

[3] Fernando, K.V., Pont, M.W., Computing Accurate Eigenvalues of a Hermitian Matrix, Diamond Doc. No. 03/3-3/1/N1.p .

[4] Gregory, R. T., Karney, D. L., A Collection of Matrices for Testing Computational Algoritms, Wiley-Interscience, New York 1969.

[5] Grüner, K., Solving Complex Linear Systems with Verified High Accuracy. Diamond Deliverable Doc. No. 03/3-1/1/K2.f, 1989.

[6] Grüner, K., Solving the Complex Eigenvalue Problem with Verified High Accuracy. Diamond Deliverable D3-1/2, Doc. No. 03/3-2/1/K1.f, 1988.

[7] Kreß, O., Über Fehlerabschätzungen für Eigenwerte und Eigenvektoren reellsymmetrischer Matrizen, University of Karlsruhe, Institut für Praktische Mathematik, Internal report Nr.73/4, 1973.

[8] McDonald, K., A Simultaneous Linear Equation Facility in Ada. Diamond Deliverable D3-1, Doc. No. 03/3-1, 3-6/1/N1.p .

[9] Moynihan, V., Techniques for Generating Accurate Eigensolutions in Ada. Diamond Doc. No. 03/3-2/1/N1.f, 1988.

[10] Rump, S. M., Kleine Fehlerschranken bei Matrixproblemen, Dissertation, Institute for Applied Mathematics, University of Karlsruhe, 1980.

[11]   Stetter, H. J., Sequential Defect Correction for High Accuracy Floating-Point Algorithms, Lecture Notes in Mathematics, Vol. 1066, p.186-202.

[12]   Stoer, J., Bulirsch, R., Introduction to Numerical Analysis. Springer, Berlin, Heidelberg, New York, 1980.

[13]   Wilkinson, J.H., The Algebraic Eigenvalue Problem. Clarendon Press, Oxford 1965.

[14]   Wilkinson, J. H., Rounding Errors in Algebraic Processes. H. M. Stationary Office, London, 1968.

[15]   Wilkinson, J. H., Reinsch, C., Linear Algebra. Handbook for Automatic Computation, Volume II, Springer, Berlin, Heidelberg, New York, 1971.

[16]   Zielke, G., Testmatrizen mit maximaler Konditionszahl, Computing 13, pp. 33-54, 1974.

[17]   Bleher, J.H., Kulisch, U., Metzger, M., Rump, S.M., Ullrich, Ch., Walter, W.: FORTRAN-SC: A Study of a FORTRAN Extension for Engineering / Scientific Computation with Access to ACRITH. Computing 39, 93-110 (1987)

[18]   Kulisch, U. (ed.): PASCAL-SC, A Pascal Extension for Scientific Computation, Information Manual and Floppy Disks. Version IBM PC/AT ; operating system DOS. Stuttgart: B. G. Teubner (Wiley-Teubner series in Computer Science) 1987.

[19]   Kulisch, U. (ed.): PASCAL-SC, A Pascal Extension for Scientific Computation, Information Manual and Floppy Disks. Version ATARI ST. Stuttgart: B. G. Teubner 1987.

Computing Accurate Eigenvalues of a Hermitian Matrix

DIAMOND deliverable D 3-1 (part)

K.V.Fernando and M.W.Pont
Numerical Algorithms Group Ltd, Wilkinson House,
Jordan Hill Road, Oxford, United Kingdom, OX2 8DR

1. Introduction

This paper will describe an algorithm to compute the eigenvalues and eigenvectors of a Hermitian matrix, with guaranteed inclusion intervals containing the eigenvalues, and will give the specification of an Ada implementation of the algorithm. The algorithm presented here incorporates a Jacobi method [5],[6],[7],[8],[10],[14] to compute a good initial approximation to the eigensolution. Newton iterations [4],[9],[11],[12] are then used to improve the eigenvalues and eigenvectors. Such improved eigenvalues are accurate to the basic machine precision.

In recent years, Jacobi methods have not received as much attention as QR algorithms, which are popular because they can generally compute an eigensolution more efficiently. Jacobi methods are still attractive, however, in terms of the accuracy of the eigensolution. The QR algorithm breaks down for problems with small off-diagonal elements [2] and hence for such problems the QR algorithm is not suitable. Furthermore, the Jacobi method gives orthogonal eigenvectors which span the correct subspace even when there are close eigenvalues (see p. 281 of [14]). Also, with the advent of parallel-processor machines, much effort has been spent in developing new Jacobi algorithms to decrease execution time and overheads – Jacobi methods are susceptible to improvement in this respect.

The problem of computing eigenvalues with guaranteed accuracy is an important one, as in certain signal processing problems where it is required to compute eigenvalues of a matrix to a high degree of accuracy. Standard methods, such as the basic QR algorithm, may return an approximation to this eigenvalue which has absolute accuracy only, with little relative accuracy if the eigenvalue is small compared to the norm of the matrix. Also, no estimate of accuracy is returned.

Once a good approximation of the eigenvalues and the corresponding eigenvectors are known (in our case from the Jacobi method), the Newton iterations can be used to improve each eigenvalue-eigenvector pair to relative machine accuracy. It is well known that the Newton iterations are similar to inverse iterations and the rate of convergence is at least quadratic under mild assumptions. The computed eigenvalue-eigenvector pair can be easily used to compute an analytical bound which determines the inclusion interval for that eigenvalue.

The rest of this paper is structured as follows:
Section 2 describes a Jacobi method for Hermitian matrices;
Section 3 illustrates how inclusion intervals for the eigenvalues may be computed;
Section 4 discusses how Newton iterations can be used to improve the eigensolution;
Section 5 discusses how the algorithm is adapted to Ada;
Section 6 gives the Ada specification of the package;
Section 7 presents certain numerical results;
Section 8 presents our conclusions.

2. A Jacobi Method for the Hermitian Eigenvalue Problem

The Jacobi method [5],[6],[7],[8],[10],[14] for computing the eigenvalue decomposition of a complex Hermitian matrix A is based on generating a sequence of matrices $\{A^{(k)}\}$ such that

$$A^{(k+1)} = J(\alpha,\beta,\gamma,\delta,\phi,i,j)^* A^{(k)} J(\alpha,\beta,\gamma,\delta,\phi,i,j) \quad , \quad A^{(0)} = A \in C^{n\times n}$$

where $J(\alpha,\beta,\gamma,\delta,\phi,i,j)$ is equivalent to an identity matrix except that the elements (i,i), (j,j), (i,j) and (j,i) are given by

$$J(\alpha,\beta,\gamma,\delta,\phi,i,j)_{i,i} = e^{j\alpha}\cos\phi$$
$$J(\alpha,\beta,\gamma,\delta,\phi,i,j)_{i,j} = e^{j\beta}\sin\phi$$
$$J(\alpha,\beta,\gamma,\delta,\phi,i,j)_{j,i} = -e^{j\gamma}\sin\phi$$
$$J(\alpha,\beta,\gamma,\delta,\phi,i,j)_{j,j} = e^{j\delta}\cos\phi$$
$$\alpha - \gamma = \beta - \delta \quad , \quad j^2 = -1 \ .$$

We have used the notation $J(\alpha,\beta,\gamma,\delta,\phi,i,j)^*$ to indicate the conjugate transpose of the matrix $J(\alpha,\beta,\gamma,\delta,\phi,i,j)$.

In our implementation we have set β and γ to zero and hence

$$\alpha = -\delta.$$

Let

$$J(\alpha,\phi) = \begin{bmatrix} e^{j\alpha}\cos\phi & \sin\phi \\ -\sin\phi & e^{-j\alpha}\cos\phi \end{bmatrix}$$

$$A_{i,j} = \begin{bmatrix} a_{i,i} & a_{i,j} \\ a_{j,i} & a_{j,j} \end{bmatrix} \quad , \quad a_{j,i} = a_{i,j}^*.$$

The rotation matrix $J(\alpha,\phi)$ is chosen such that the elements (i,j) and (j,i) of the matrix $A^{(k)}$ are annihilated. That is

$$J(\alpha,\phi)^* A_{i,j} J(\alpha,\phi) = diag(\hat{a}_{i,i},\hat{a}_{j,j}).$$

To compute the eigenvector matrices, the following sequence is also generated.

$$U^{(k+1)} = U^{(k)} J(\alpha,\phi,i,j) \quad , \quad U^{(0)} = I$$

If a Jacobi algorithm is convergent, then

$$A^{(k)} \to \Lambda \quad , \quad U^{(k)} \to U \quad as \quad k \to \infty$$

where

$$A = U\Lambda U^*$$

is the eigenvalue decomposition of the matrix.

Perhaps, the best routine available in the literature for the Jacobi problem is due to Rutishauser [10] which is written in Algol 60. However, there are a few shortcomings in this routine. Particularly,

(a) The Rutishauser routine is for the symmetric problem and not for the Hermitian problem.

(b) It does not exploit the symmetry (Hermitian structure) of the problem which can be used to reduce the complexity of the problem.

(c) The four elements of the matrix $A_{i,j}$ which are required to compute the rotation necessary to annihilate $a_{i,j}$ and $a_{j,i}$ are not (in general) in contiguous positions of the matrix A. These four elements require updating after computation of the rotation.

Also, the rows (columns) i and j of the matrix A which get affected by this rotation are, in general, not in contiguous positions. Thus, in paged machines, the Rutishauser routine will be inefficient from the point of view of page faults.

It is not too difficult to extend the Jacobi algorithm to the Hermitian case. To overcome the problems (b) and (c), we have decided to implement a modified row cyclic Jacobi algorithm.

In our algorithm, we copy the diagonal elements of the matrix A to a one dimensional array and the computations are done using this array and the strictly lower part of matrix A. If the off-diagonal element $a_{i,j}$ $i > j$ is available, then the element $a_{j,i}$ is implicitly known and hence it is not necessary to use the upper part of the matrix in the algorithm.

In the usual row cyclic Jacobi method which is also known as the special cyclic method [14], the indices (i,j) are chosen in the following order:

$$(1,2),(1,3),...,(1,n); \quad (2,3),...,(2,n); \quad ... \; ; \quad (n-1,n)$$

where n denotes the order of the square matrix. This requires diagonalisation of N 2×2 matrices $A_{i,j}$ which constitute a (nominal) sweep where $N = n(n-1)/2$. Sweeps are repeated until the matrix $A^{(k)}$ becomes diagonal.

In the modified version, the annihilations are along the first sub-diagonal in the following manner:

$$(1,2),(2,3),...(n-1,n); \quad (1,2),(2,3),...(n-2,n-1); \quad ... \; ; \quad (1,2)$$

This ordering was first suggested by Gentleman (see [5]).

However, it is not necessary to do the last annihilation corresponding to the pivot (1,2) as it will be done in the next sweep. Thus, this method has only $N-1$ annihilations in a sweep.

If $a_{i,j}$ is not already equal to zero, then the angle ϕ can be chosen in two ways. In standard implementations of the Jacobi method, the solution

$$|\phi| \leq \pi/4,$$

known as an inner rotation, is implemented. However, in our design, we have used the outer rotational solution,

$$\pi/2 \geq |\phi| \geq \pi/4.$$

If we use outer rotations for annihilations in the modified method, then it is equivalent to the standard row cyclic method with inner rotations and hence the linear (global) and quadratic convergence properties of these methods are equivalent [5].

It is easily seen that the pivots are contiguous and hence the first problem mentioned in (c) is absent in the modified method. Although it is not a requirement in standard Ada [1], most compilers are designed such that the elements of two dimensional arrays are stored row-wise. In that case, application of the pre-rotation $J(\alpha,\phi)^*$ will not be a problem in paged machines.

If we apply the post-rotation $J(\alpha,\phi)$ after the annihilation of the off-diagonal elements of the matrix $A_{i,j}$, then the elements $a(l,i)$, $a(l,j)$, $l = j+1,...,n$ will be affected and this will lead to 'page thrashing' for sufficiently large n. However, it is not necessary to apply the rotation immediately after this annihilation to all the elements $a(l,i)$, $a(l,j)$, $l = j+1,...,n$ as they can be delayed until it is essential to apply them. However, before the next annihilation (of the matrix $A_{j,j+1}$) we have to apply this post-rotation to the two elements $a(j+1,i)$ and $a(j+1,j)$. In the same vein, if we have delayed the application of the pre-rotations computed using the 2×2 matrices $A_{l,l+1}$, $l = 1,...,i$, to the elements $a(j+1,l)$, $a(j+1,l+1)$, $l =1,...,i$, respectively, then they can be applied before the $A_{j,j+1}$ annihilation. Thus, by delaying the post-rotations, it is possible to transform them into row-wise operations which will remove the problem of 'page thrashing'.

3. Inclusion of the Estimated Eigenvalues

Once an eigenvalue is computed, it is often required to know the *a posteriori* accuracy of this estimated eigenvalue. The following analysis is due to Wilkinson [13].

Suppose that we have computed an alleged eigenvalue λ and the corresponding alleged eigenvector x of the Hermitian matrix A. Let v be the vector,

$$v = Ax - \lambda x \tag{3.1}$$

which may be called the residual vector corresponding to the alleged eigensolution (λ, x) where we have assumed that the vector x is normal,

$$\|x\|_2 = 1 .$$

Since A is Hermitian, there exists a unitary matrix U such that

$$UAU^* = diag\{\lambda_1, \ldots, \lambda_n\}$$

where λ_i, $i = 1, \ldots, n$ are the eigenvalues of the matrix.

Hence, (3.1) gives,

$$diag\{(\lambda_1 - \lambda), \ldots, (\lambda_n - \lambda)\}Ux = Uv$$

$$Ux = diag\{(\lambda_1 - \lambda)^{-1}, \ldots, (\lambda_n - \lambda)^{-1}\}Uv$$

$$\|Ux\|_2 \leq \|diag\{(\lambda_1 - \lambda)^{-1}, \ldots, (\lambda_n - \lambda)^{-1}\}\|_2 \|Uv\|_2$$

$$1 \leq \max_i |\lambda_i - \lambda|^{-1}\Delta$$

where

$$\Delta = \|v\|_2.$$

Then,

$$\min_i |\lambda_i - \lambda| \leq \Delta$$

which indicates that there exists at least one eigenvalue in the closed interval $[\lambda - \Delta, \lambda + \Delta]$.

4. Improvement of the Eigensolution by Newton Iterations

One of the most interesting algorithms available to improve the approximate eigensolution of a Hermitian matrix is the method of Newton iterations [4],[9],[11],[12]. The convergence rate of this method is at least quadratic (under mild assumptions) and often one iteration is adequate to obtain a very accurate eigenvalue and the corresponding eigenvector.

Although, this is a powerful method, the derivation of the algorithm is quite elementary. In finding a distinct eigenvalue λ and the corresponding normalized (with respect to $\|.\|_2$) eigenvector x, we require the solution of the nonlinear system

$$Ax - \lambda x = 0$$

$$x^*x = 1$$

If $(\lambda^{(k)}, x_{(k)})$ is an eigenpair (with $x_{(k)}^* x_{(k)} = 1$), the true pair $(\lambda^{(k)} + \delta\lambda^{(k)}, x_{(k)} + \delta x_{(k)})$ satisfies exactly the relations,

$$A(x_{(k)} + \delta x_{(k)}) - (\lambda^{(k)} + \delta\lambda^{(k)})(x_{(k)} + \delta x_{(k)}) = 0$$

$$(x_{(k)} + \delta x_{(k)})^*(x_{(k)} + \delta x_{(k)}) = 1 .$$

If we ignore the second order quantities, we get

$$(Ax_{(k)} - \lambda^{(k)}x_{(k)}) + (A - \lambda^{(k)}I)\delta x_{(k)} - \delta\lambda^{(k)}x_{(k)} = 0$$

$$x_{(k)}^* \delta x_{(k)} = 0 .$$

The above two equations give the system

$$\begin{bmatrix} (A-\lambda^{(k)}I) & x_{(k)} \\ x_{(k)}^* & 0 \end{bmatrix} \begin{bmatrix} \delta x_{(k)} \\ -\delta\lambda^{(k)} \end{bmatrix} = \begin{bmatrix} -v_{(k)} \\ 0 \end{bmatrix} \tag{4.1}$$

where the residual $v_{(k)}$ is defined as

$$v_{(k)} = Ax_{(k)} - \lambda^{(k)}x_{(k)} \quad .$$

Thus, by solving an $(n+1)\times(n+1)$ system of equations, we, can obtain an improved eigensolution. We emphasise that it is not necessary to solve this set of equations (4.1) in double precision and in our implementation we have used a Crout reduction type technique. However, there are more efficient algorithms (see Chapter 5 of [3]) which exploit the Hermitian structure of the system of linear equations (4.1).

In our implementation of the Newton algorithm, we have used double precision arithmetic to compute the residual $v_{(k)}$. The approximate eigensolution $(\lambda^{(0)}, x_{(0)})$ which is necessary to initiate the Newton algorithm is obtained from the Jacobi method (see Section 2).

There are many ways to terminate the algorithm and we used the following criterion.

$$\Delta^{(k)} < \varepsilon\lambda^{(k)}$$

where

$$\Delta^{(k)} = \|v_{(k)}\|_2$$

and ε is the basic machine precision.

In view of the results given in Section 3, there exists an eigenvalue in the open interval $\left(\lambda^{(\bar{k})}(1-\varepsilon), \lambda^{(\bar{k})}(1+\varepsilon)\right)$ where $\bar{k}$ is the the final value of k. Thus, it is reasonable to state that the computed eigenvalue $\lambda^{(\bar{k})}$ is at least accurate to the basic machine precision.

The improved eigenvalue $\lambda^{(k)}$ is obviously given by '

$$\lambda^{(k+1)} = \lambda^{(k)} + \delta\lambda^{(k)}. \tag{4.2}$$

The quantities $\lambda^{(k)}$ and $\delta\lambda^{(k)}$ are basic precision real numbers in our implementation. However, if these real values are added together in double precision, then the accuracy of the result can exceed basic precision. Particularly, the final result $\lambda^{(\bar{k})}$ can be obtained via (4.2) with double precision arithmetic and hence the accuracy of this eigenvalue can approach double precision accuracy.

In our analysis and implementation, we have assumed that the eigenvalues of the Hermitian matrix A are distinct. The Newton method as stated (in general) is not suitable for matrices with non-distinct eigenvalues. However, it is possible to extend the basic Newton method to such cases and we hope to implement such algorithms in the near future.

The Newton algorithm can also break down if the matrix has eigenvalues with magnitudes approximately equal to $\varepsilon\|A\|$. In that case, it might not be possible to compute such eigenvalues accurately without non-trivial amounts of computation in double precision arithmetic.

5. Adapting the Algorithm to Ada

It was desired to create a generic package which could be instantiated by the user for use with either real symmetric or complex Hermitian matrices, both of which have real eigenvalues. The Jacobi algorithm for the two matrix types is sufficiently similar that very little inefficiency is introduced by having one generic package.

Since the matrices are Hermitian, we clearly need only $n(n+1)/2$ elements of storage for a matrix of order n; however, it was decided not to require that matrices should be stored in packed form, but to use n^2 storage elements per matrix. The package then becomes slightly simpler to write, but also, since we need the original matrix elements to compute the bounds on the eigenvalues, we are saved the need of making a copy of the matrix before applying the Jacobi algorithm, which overwrites the matrix elements – we already have a copy in the opposite triangle of the matrix.

The package is thus generic with respect to the following types:

type FLOAT_TYPE, which is a real type of any precision;
type FLOAT_VECTOR_TYPE, a vector of FLOAT_TYPE;
type SCALAR_TYPE, which is a private real or complex type, and will generally have the same precision as FLOAT_TYPE;
type VECTOR_TYPE, a vector of SCALAR_TYPE;
type MATRIX_TYPE, a vector of VECTOR_TYPE;

In addition, we need to allow the user some generality in how he stores his matrices. We should not force him to store matrices with element $(1,1)$, say, at the top left-hand corner. In any case, for some purposes he may wish to operate on sub-matrices of a larger matrix.

We therefore use a NAG Ada Library defined type TRIM, which specifies the lower and upper row and column bounds of a sub-matrix. The user can alter these bounds according to which sub-matrix he wishes to access. With both the input data matrix and the output eigenvector matrix is associated a variable of type TRIM (obtained from NAG_A01BA in the NAG Ada Library).

While the storage order of two-dimensional arrays is not specified by the Ada Language Reference Manual [1], all compilers known to the authors store them in row major order. With this in mind, the algorithm was coded in Ada to operate row-wise on the data matrix for efficiency, and to store the returned eigenvectors row-wise.

As described in Section 4, in order to give inclusion intervals for the eigenvalues we need to compute the quantities $\|Ax_i - \lambda_i x_i\|_2$, where λ_i is an eigenvalue of A and x_i is the corresponding eigenvector. It is left to the user to supply a function that returns these values to the package – accurate arithmetic should be used to implement this function. In the package specification (Section 6) this function is called ACCURATE_RESIDUAL_NORM.

The user must also pass to the generic package functions for arithmetic operations on the types FLOAT_TYPE and SCALAR_TYPE, functions for converting from one type to the other, and such things as square-root, absolute value and complex conjugate functions. When instantiating with a real type for SCALAR_TYPE, some of these functions will need to do nothing but return their input argument. This should not be inefficient if pragma INLINE is used.

Other objects used from the NAG Ada Library are the error handling mechanism, and the exception LIBRARY_EXCEPTION from NAG_P01AA.

6. Ada Package Specification

Procedure ROWWISE_SYMMETRIC_JACOBI is the only user-callable routine in the package; it takes as input the symmetric matrix A and the TRIM type A_TRIM describing which sub-matrix of A to operate on, and returns the eigenvalues of the sub-matrix in argument EIGENVALUES.

Arguments COMPUTE_EIGENVECTORS and COMPUTE_EIGENVALUE_BOUNDS, both BOOLEAN, allow the user to specify whether or not he also wishes to compute the eigenvectors of the sub-matrix, and the inclusion intervals of the eigenvalues, respectively. If the eigenvectors are required, a matrix must be supplied for them as argument EIGENVECTORS, with the TRIM type EIGENVECTORS_TRIM to say which sub-matrix of EIGENVECTORS the results should be stored in. The eigenvectors are stored in the rows of this sub-matrix rather than the columns, for efficiency.

The user must also specify the maximum number of sweeps of the Jacobi algorithm he wishes to allow, in INTEGER argument ITERATIONS. On exit, this argument contains the actual number of sweeps taken.

Note that the eigenvectors are required in order to compute the eigenvalue bounds; therefore COMPUTE_EIGENVALUE_BOUNDS cannot be TRUE unless COMPUTE_EIGENVECTORS is also TRUE.

Procedure ROWWISE_SYMMETRIC_JACOBI raises the NAG library exception LIBRARY_EXCEPTION if any of the following conditions occur:

(i) The sub-matrix described by A_TRIM is incompatible with matrix A, or A_TRIM describes a sub-matrix of order less than 1, or A_TRIM does not describe a square sub-matrix.

(ii) Argument COMPUTE_EIGENVECTORS is TRUE, and the sub-matrix described by EIGENVECTORS_TRIM is incompatible with matrix EIGENVECTORS, or A_TRIM is incompatible with EIGENVECTORS_TRIM.

(iii) Vector EIGENVALUES has length less than N, where N is the order of the sub-matrix defined by A_TRIM.

(iv) COMPUTE_EIGENVALUE_BOUNDS is TRUE, and vector EIGENVALUE_BOUNDS has length less than N described above.

(v) The Jacobi algorithm takes more sweeps to complete than the upper limit specified by the user in argument ITERATIONS.

In all of the above cases, information about the error is written to argument FAIL of the NAG library type ERROR_RECORD, and this can be examined by the user.

We now present the specification of the Ada package.

```
with NAG_A01BA; use NAG_A01BA;        -- Defines type TRIM for submatrices.
with NAG_P01AA; use NAG_P01AA;        -- NAG library error mechanism.
with NAG_P01AD;                       -- NAG library exception definitions.
generic
  type FLOAT_TYPE is digits <>;       -- A real type that defines the
                                      -- precision of the matrix.

  type FLOAT_VECTOR_TYPE is array (INTEGER range <>) of FLOAT_TYPE;

  type SCALAR_TYPE is private;         -- Can be either a real type or
                                      -- a complex type, but should have
                                      -- the same precision as FLOAT_TYPE.

  type VECTOR_TYPE is array (INTEGER range <>) of SCALAR_TYPE;

  type MATRIX_TYPE is array (INTEGER range <>, INTEGER range <>) of
                                                           SCALAR_TYPE;
  ZERO_SCALAR : SCALAR_TYPE;           -- Real or complex 0.0
  UNIT_SCALAR : SCALAR_TYPE;           -- Real or complex 1.0
  with function SIGN (X : FLOAT_TYPE; Y : FLOAT_TYPE := 1.0)
                                       return FLOAT_TYPE is <>;
       -- Returns the second argument with the sign of the first.
  with function SQRT (X : FLOAT_TYPE) return FLOAT_TYPE is <>;
       -- Returns the square root of X.
  with function SCALAR_TO_FLOAT(X : SCALAR_TYPE) return FLOAT_TYPE is <>;
       -- Converts from SCALAR_TYPE to FLOAT_TYPE; may be a null function
       -- if SCALAR_TYPE is the same as FLOAT_TYPE.
  with function FLOAT_TO_SCALAR(X : FLOAT_TYPE) return SCALAR_TYPE is <>;
       -- Converts from FLOAT_TYPE to SCALAR_TYPE; may be a null function
       -- if SCALAR_TYPE is the same as FLOAT_TYPE.
```

```
   with function CONJG (X : SCALAR_TYPE) return SCALAR_TYPE is <>;
           -- Returns the complex conjugate of X; may be a null function if
           -- SCALAR_TYPE is a real type.
   with function ACCURATE_RESIDUAL_NORM (A : in MATRIX_TYPE;
                                A_TRIM : in TRIM;
                                     X : in VECTOR_TYPE;
                                LAMBDA : in FLOAT_TYPE)
                                              return FLOAT_TYPE is <>;
           -- Computes the value  2_norm(A*X-LAMBDA*X) accurately, for
           -- the eigenvalue inclusion interval.
   with function "abs" (X : FLOAT_TYPE) return FLOAT_TYPE is <>;
           -- returns the absolute value of X.
   with function "abs" (X : SCALAR_TYPE) return FLOAT_TYPE is <>;
           -- returns the absolute value of X. May be the same function as
           -- above if SCALAR_TYPE is the same as FLOAT_TYPE.
   with function "*" (X : FLOAT_TYPE; Y : SCALAR_TYPE)
                                              return SCALAR_TYPE is <>;
   with function "/" (X : SCALAR_TYPE; Y : FLOAT_TYPE)
                                              return SCALAR_TYPE is <>;
   with function "-" (X : SCALAR_TYPE) return SCALAR_TYPE is <>;
   with function "+" (X, Y : SCALAR_TYPE) return SCALAR_TYPE is <>;
   with function "-" (X, Y : SCALAR_TYPE) return SCALAR_TYPE is <>;
   with function "*" (X, Y : SCALAR_TYPE) return SCALAR_TYPE is <>;
   with function "/" (X, Y : SCALAR_TYPE) return SCALAR_TYPE is <>;


   package JACOBI_GENERIC is
     procedure ROWWISE_SYMMETRIC_JACOBI(A : in out MATRIX_TYPE;
                              A_TRIM : in TRIM;
                         EIGENVALUES : in out FLOAT_VECTOR_TYPE;
            COMPUTE_EIGENVALUE_BOUNDS : in BOOLEAN;
                    EIGENVALUE_BOUNDS : in out FLOAT_VECTOR_TYPE;
                 COMPUTE_EIGENVECTORS : in BOOLEAN;
                        EIGENVECTORS : in out MATRIX_TYPE;
                    EIGENVECTORS_TRIM : in TRIM;
                          ITERATIONS : in out INTEGER;
                                FAIL : in ERROR_RECORD := DEFAULT_RECORD);
   end JACOBI_GENERIC;
```

7. Test Results

The method described in Sections 2 to 4 was applied to several test matrices. We give now the results obtained for three tests, using DEC VAX double precision (G_Floating) arithmetic as basic precision. Thus, in our machine, a floating-point number is stored in base 2, and has 53 mantissa bits.

For the first test example, we have chosen a well-conditioned problem, while the second and third examples use Hilbert matrices, which are very ill-conditioned.

Test 1

Random matrix of order 11, with each element chosen uniformly from the interval [−1.0,1.0].

i	λ_i	Δ_i	iterations	t_1	t_2
1	−2.9779680972461637E+00	6.31E−31	1	102	103
2	−2.7500948041107520E+00	5.63E−31	1	102	103
3	−1.9911320360037315E+00	5.52E−31	1	102	103
4	−1.6590377721176464E+00	1.26E−30	1	100	102
5	−6.6030579519503818E−01	2.89E−31	1	101	104
6	−2.2170114573794680E−01	2.65E−31	1	99	104
7	3.3710215454392267E−01	4.36E−31	1	99	104
8	1.1208063521479070E+00	7.46E−31	1	100	103
9	1.8165696767486637E+00	3.83E−31	1	102	104
10	2.1472345912681989E+00	3.53E−31	1	102	104
11	3.0278070765560341E+00	3.11E−31	1	103	104

In the table above, the eigenvalues λ_i are arranged in ascending order. The inclusion interval for each eigenvalue is specified by Δ_i; thus for each i, the true eigenvalue λ_{true} lies between $\lambda_i - \Delta_i$ and $\lambda_i + \Delta_i$. This bound has to be computed with accurate arithmetic. (Note that not all the significant figures of λ_i are necessarily printed in the table, and that λ_i may be represented to double precision by two basic precision numbers).

The column headed **iterations** denotes the number of inverse iterations performed to improve each inclusion interval.

Finally, the columns headed by t_1 and t_2 contain two measures of the accuracy of the eigenvalue λ_i, the first being relative precision, and the second absolute precision. The measures are defined as

$$t_1 = int\{-\log_2(\Delta_i/|\lambda_i|)\},$$

$$t_2 = int\{-\log_2(\Delta_i/\|A\|_F)\},$$

where $\|\cdot\|_F$ denotes the Frobenius norm. These two measures show approximately how many bits of accuracy each eigenvalue has, relative to the size of the eigenvalue and the norm of the matrix, respectively.

Random matrices such as the above will in general be well conditioned. For all such matrices tested, the number of steps required for convergence of the inverse iteration technique was 1, for all eigenvalues. Although double precision arithmetic is only used to compute the residual v, our results are all accurate almost to double precision.

Test 2

Hilbert matrix of order 11, scaled so that the matrix is exactly storable: scaling factor = 232792560.

i	λ_i	Δ_i	iterations	t_1	t_2
1	7.8991604348341898E−07	2.30E−23	3	55	104
2	1.8174298649650591E−04	2.64E−21	3	56	97
3	1.9282973274184198E−02	4.87E−19	2	55	89
4	1.2497185430266273E+00	3.79E−17	2	55	83
5	5.5212795964291978E+01	2.00E−18	1	65	87
6	1.7558159828273706E+03	5.66E−20	1	75	93
7	4.1304037745128713E+04	6.31E−21	1	82	96
8	7.2501714008627227E+05	7.44E−23	1	93	102
9	9.3837800294171870E+06	4.33E−23	1	97	103
10	8.4359863438188002E+07	1.68E−23	1	102	104
11	4.1317959905660057E+08	5.56E−24	1	106	106

The table column headings are as described for test 1. In this test, we see that for matrices with eigenvalues that differ greatly in magnitude, the number of inverse iterations needed for convergence of the smaller eigenvalues increases.

Test 3

Hilbert matrix of order 18, scaled so that the matrix is exactly storable: scaling factor = 144403552893600.

i	λ_i	Δ_i	iterations	t_1	t_2
1	−1.4870947057432651E−04	7.74E−04	14		58
2	−1.9278665876242651E−05	1.19E−04	10		61
3	2.2733773888017248E−03	4.40E−02	0		53
4	2.8553616692135152E−03	2.29E−02	0		53
5	8.7224948553915188E−03	1.14E−19	23	56	111
6	5.5430398847354545E−01	5.31E−18	7	57	105
7	2.7729282467290627E+01	1.24E−15	4	54	98
8	1.1194809886034511E+03	1.04E−13	2	53	91
9	3.7092511702932316E+04	3.20E−12	2	53	86
10	1.0201140010530915E+06	1.01E−11	2	56	85
11	2.3451737223474666E+07	1.69E−12	1	64	87
12	4.5223426995204306E+08	3.61E−14	1	73	93
13	7.3138397012674112E+09	8.25E−15	1	80	95
14	9.8759016585691025E+10	4.89E−16	1	87	99
15	1.1013017772443303E+12	3.86E−17	1	95	103
16	9.9016143067941992E+12	1.40E−17	1	99	104
17	6.7126846443011008E+13	1.53E−17	1	102	104
18	2.7223185553476603E+14	8.13E−18	1	105	105

Where column t_1 is left blank, the inverse iteration process has failed to converge – this may happen as described in Section 4. However the computed eigenvalues still have a guaranteed inclusion.

From the table it can be seen that the four smallest eigenvalues failed to converge with inverse iteration – indeed, since Hilbert matrices are positive-definite, the two smallest eigenvalues have the wrong sign.

Eigenvalue λ_5 took 23 iterations to converge – for Hilbert matrix problems, the smallest eigenvalue that does converge takes the largest number of iterations. This eigenvalue has size approximately ε times $\|A\|_F$ where ε is the relative machine precision (approximately 2^{-53} in G_Floating arithmetic).

When the Jacobi part of the algorithm terminates with several eigenvalue approximations having the same order of magnitude, as in this case, it is possible that the inverse iteration technique using one eigenvalue-eigenvector pair as a starting point will converge onto a different eigenpair. A simple check on this is to compute the value $|x^*y|$, where x is the eigenvector corresponding to the eigenvalue initially and y is the eigenvector corresponding to the improved eigenvalue. Since the eigenvectors are theoretically orthonormal this quantity should be approximately unity. If the method converges on to a different eigenpair, the quantity will be near zero.

8. Conclusions

We have developed a Jacobi algorithm to obtain eigenvalues and eigenvectors of complex Hermitian or real symmetric matrices. We have also designed a Newton type iterative algorithm to improve the eigensolutions given by the Jacobi method if further accuracy of the solution is desired. This Newton step gives eigenvalues which are accurate to relative machine precision.

These algorithms have been implemented in Ada and have been thoroughly tested and appear to be robust. Since the accuracy of each eigenvalue is available from the Newton algorithm, our implementation is self-testing. Although such self-testing algorithms can be expensive for ill-conditioned problems, we believe that such methods are indispensible if high quality results are required.

9. References

[1]   American National Standards Institute, Inc.
       Reference Manual for the Ada® Programming Language.
       ANSI/MIL-STD-1815 A- 1983, 1983.

[2]   Blevins, M.M. and Stewart, G.W.
       'Calculating the eigenvectors of diagonally dominant matrices'.
       J. ACM, 21, pp. 261-271, 1974.

[3]   Dongarra, J.J., Moler, C.B., Bunch, J.R. and Stewart, G.W.
       Linpack User's Guide.
       SIAM, Philadelphia, 1979.

[4]   Dongarra, J.J., Moler, C.B. and Wilkinson, J.H.
       'Improving the accuracy of computed eigenvalues and eigenvectors'.
       SIAM J. Numer. Anal., 20, pp. 23-45, 1984.

[5]   Fernando, K.V.
       'Linear convergence of the row cyclic Jacobi and Kogbetliantz methods'.
       NAG Technical Report, TR5A/88, Numerical Algorithms Group Ltd, Oxford, 1988.

[6]   Jacobi, C.J.G.
       'Über eine neue Auflösungsart der bei Methode der Kleinsten Quadrate vorkommenden
       lineären Gleichungen'.
       Astronomische Nachrichten, 22, pp. 297-306, 1845.

[7]   Jacobi, C.J.G.
        'Über ein leichtes Verfahren die in der Theorie der Säcularstörungen vorkommenden
        Gleichungen numerisch aufzulösen'.
        Journal reine angew. Mathematik (Crelle's Journal), 30, pp. 51-94, 1846.

[8]   Parlett, B.N.
        The Symmetric Eigenvalue Problem.
        Prentice Hall, Englewood Cliffs, NJ, 1980.

[9]   Peters, G. and Wilkinson, J.H.
        'Inverse iteration, ill-conditioned equations and Newton's method'.
        SIAM Review, 21, pp. 339-360, 1979.

[10]   Rutishauser, H.
        'The Jacobi method for real symmetric matrices'.
        Numer. Math., 9, pp. 1-10, 1966.

[11]   Santos, M.C.
        'A note on the Newton iteration for the algebraic eigenvalue problem'.
        SIAM J. Matrix Anal., 9, pp. 561-569, 1988.

[12]   Symm, H.J. and Wilkinson, J.H.
        'Realistic error bounds for a simple eigenvalue and its associated eigenvector'.
        Numer. Math., 35, pp. 113-126, 1980.

[13]   Wilkinson, J.H.
        Rounding Errors in Algebraic Processes.
        HMSO, London, 1963.

[14]   Wilkinson, J.H.
        The Algebraic Eigenvalue Problem.
        Clarendon Press, Oxford, 1965.

Verified Inclusion of all Roots of a Complex Polynomial
by means of Circular Arithmetic

Diamond Deliverable D3-2

W. Frangen
Universität Karlsruhe

Abstract:

This report concerns the implementation of a method for locating all roots of a complex polynomial in the complex plane. This method has been described by P. Henrici (1974). It essentially consists of bisecting the complex plane by squares and then testing a square to see if it is free of roots. "In this decision, the algorithm (excepting rounding errors) never fails." This sentence was written by P. Henrici in a previous publication, see Dejon/Henrici (1969), p.99.

For a strict verification, the drawback of rounding errors needs to be overcome by a sort of interval arithmetic. This is feasible

1. theoretically, by the Kulisch arithmetic, see Kulisch/Miranker (1981), (1983);

2. practically, by a programming language providing the directed roundings according to the principles of the Kulisch arithmetic. In fact, I used the PASCAL-SC language, see Bohlender/Rall/Ullrich/Wolff v. Gudenberg (1986);

3. finally, by means of a suitable computer. I used a SAM 68K. Its decimal floating point numbers have a 13 place mantissa and a decimal exponent reaching from -99 to +99.

The implementation has been confronted with diverse problems not dealt with in the literature. These will form the main part of my report.

1. Introduction

1.1 Problem

A complex polynomial of degree n (n=0, 1, ...) has the form

$$p(z) = p_n z^n + p_{n-1} z^{n-1} + \ldots + p_1 z + p_0,$$

$$p_n \neq 0,$$

with complex coefficients p_i (i=0, ..., n). According to the fundamental theorem of algebra, there are exactly n roots, w_i, of the polynomial so that

$$p(z) = \prod_{i=1}^{n} (z - w_i).$$

The main difficulty in solving the polynomial equation p(z)=0 consists in the multiple roots $w_i = w_j$, $i \neq j$. They cannot be attacked by the usual Newton methods. There are algebraical methods that exactly replace the multiple roots by simple roots, by dividing p(z) by the greatest common divisor of p(z) and its derivative p'(z); see Collins (1967). Henrici's method, however, is not algebraical but analytical; it cannot distinguish between multiple roots and a cluster of roots contained in a small open disk of the complex plane.

It is well known that the roots of a polynomial may be very sensitive to minimal changes of a coefficient; see Wilkinson (1959), p. 152. Thus, for a strict verification of the roots it is not allowed to change the coefficients other than to scale all the coefficients or all the roots by a power of the base 10.

1.2 Schur/Cohn algorithm

The main tool of Henrici's method is the Schur/Cohn algorithm. If the degree and the coefficients of the polynomial are given, it delivers the number of roots z with |z|<1 (roots lying in the open unit disk).

The number of roots in an arbitrary open disk with center z_0 and radius ρ can be computed by means of a polynomial transformation $r(Z) = p(z_0 + \rho Z)$. Indeed, for each root z of $p(z)$ with $|z - z_0| < \rho$ there is a root Z of $r(Z)$ with $|Z| < 1$, and vice versa. The transformation is done in two steps:

    1. Shift: $q(z) = p(z_0 + z)$,

    2. Scaling: $r(z) = q(\rho z)$.

The shift is computed in the usual way by the complete Horner scheme for the Taylor formula. This does not cause special problems. The scaling, however, will require precautions against overflow and, as far as possible, against underflow, since the coefficients of r are $r_i = q_i \rho^i$, and the factors ρ^n of the highest coefficient may assume positive values $\ll 1$.

Of course in interval arithmetic an underflow will be replaced by an interval containing zero. This means that this interval and other intervals produced by multiplication with it will be heavily inflated. This should be avoided as far as possible.

The Schur/Cohn algorithm itself runs as follows. Let $p(z)$ be any complex polynomial of degree $\leq m$. Remark that, for our purposes, the degree m is also ascribed to the polynomial $p(z)$ if its real degree is less than m. The Schur transform $\tilde{p}$ of p is the polynomial whose coefficients are

$$\tilde{p}_j = \bar{p}_0 p_j - p_m \bar{p}_{m-j} \qquad (j = 0, \ldots, m).$$

The degree of this polynomial is $\leq m-1$ since $\tilde{p}_m = 0$. To the Schur transform $\tilde{p}$ the degree $m-1$ is ascribed. The constant part $\tilde{p}_0$ of the Schur transform is a pure real number since

$$\tilde{p}_0 = \bar{p}_0 p_0 - p_m \bar{p}_m = |p_0|^2 - |p_m|^2.$$

We begin with a polynomial $p^{(0)}(z) = p(z)$ of degree n and apply the Schur transformation as often as possible. The k-th Schur transform of p is a polynomial $p^{(k)}$ to whom the degree $n-k$ is ascribed. Now the constant parts of the successive Schur transforms will serve as (real) test numbers:

$$\gamma_k = p_0^{(k)} \qquad (k=1,\ldots,n).$$

Theorem 1 (Schur/Cohn/Henrici). Let p be a polynomial of degree n. Let all test numbers $\gamma_k \neq 0$ $(k=1,\ldots,m)$. If those indices k for which $\gamma_k < 0$ are denoted by k_j $(j=1,\ldots,m)$, where $k_1 < k_2 < \ldots < k_m$, then the number $h(p)$ of roots w of p satisfying $|w| < 1$ (multiple roots counted with their multiplicity) is given by

$$h(p) = \sum_{j=1}^{m} (-1)^{j-1} (n+1-k_j).$$

Example 1. For n=9, let the test numbers

γ_k =	9.3,	-3.1,	5.0,	2.7	-5.2,	1.3,	-1.7,	4.8,	-2.5
for k =	1	2,	3	4,	5,	6,	7,	8,	9.
		=			=		=		=
Then		k_1,			k_2,		k_3,		k_4;
$(n+1-k_j)$ =		8,			5,		3,		1;
$h(p)$ =		8 - 5 + 3 - 1 = 5.							

As P. Henrici (1974), p.496, points out, the number $h(p)$ of roots in the open unit disk remains <u>uncertain</u> if some γ_k is zero. The Schur/Cohn algorithm then is broken off.

1.3 Schur test

Under all circumstances, the Schur/Cohn algorithm can be used to decide whether a given polynomial is <u>free of roots</u> in the closed unit disk. For this aim, the algorithm can be simplified to the <u>Schur test</u>, breaking off the algorithm as soon as some test number <0 or =0 appears; otherwise all test numbers are calculated and >0. In this latter case, p has no root in the closed unit disk. If, however, the Schur test ends with a test number <0, then p has at least one root in the open unit disk. In this case, time is spared compared with the full Schur/Cohn algorithm. The Schur test is sufficient to <u>exclude</u> regions of the complex plane containing no roots of p so that the roots can be included in smaller and smaller remaining areas.

In the case of uncertainty (some $\gamma_k=0$) it is possible that some roots are lying exactly <u>on</u> the unit circle; but this is not the only possiblity; see Henrici (1974), p.496. In this case the closed unit disk is <u>suspected</u> to contain roots of the polynomial, but it is not sure if it really does.

1.4 Henrici's bisecting method

Now Henrici's <u>bisecting method</u> for including all roots w_i of the polynomial p can be described roughly; see Henrici (1974), p.513-522. It begins with an open disk containing all the roots. The radius of such a disk is

$$\sigma_0 = 2 \cdot \max_{i=1\cdots n} |p_{n-i}|^{1/i} .$$

This disk is included in a square with centre 0 and semidiagonal $\sigma_0\sqrt{2}$. The square is subdivided into four subsquares with half the diagonal of the given square. (In the following, "subsquare" always will mean one of the four subsquares which are produced in this manner). Applying the Schur test to each of these subsquares may bring out that some of the subsquares are free of roots. The remaining subsquares generally are <u>suspected</u> to contain roots; together they form a set S_{-1}.

The step no. k (k = 0,1,...) of the bisecting process will produce a set S_k consisting of those subsquares of the squares of S_{k-1} that remain after removing all subsquares that are free of roots according to the Schur test. The bisecting process will end with some set S_K. Now a final step will use the full Schur/Cohn algorithm in order to find out the exact number of roots contained in the minimal open disks covering the squares of S_K.

1.5 A cubically convergent Newton method

Henrici (1974), p.534-537, developed a cubically convergent Newton method that will refine the inclusions of polynomial roots on the following premises:

(1) All roots of the polynomial are simple and included in non-overlapping disks W_i (i=1,...,n);

(2) $\rho/\epsilon \geq 6(n-1)$, where ϵ is the maximum radius of the disks W_i and ρ is the minimum distance of the center of any disk from any of the remaining disks.

The algorithm runs as follows.
For m = 0,1,2,... let for k = 1,2,...,n

$$z_k^{(m)} := \text{mid } W_k^{(m)} \qquad \text{(middle of the circle } W_k^{(m)}),$$

$$V_k^{(m)} := \sum_{\substack{j=1 \\ j \neq k}}^{n} \frac{1}{z_k^{(m)} - w_j^{(m)}} \qquad \text{(auxiliary circles)},$$

$$W_k^{(m+1)} := z_k^{(m)} - \frac{1}{q(z_k^{(m)}) - V_k^{(m)}} \qquad \text{(new circles)},$$

where $q(z) := p'(z)/p(z)$ is the logarithmic derivative of $p(z)$.

__Theorem 2__ (Henrici). Let $W_k^{(0)} = W_k$. Then the root w_k is contained in all circles $W_k^{(m)}$ (m = 1,2,...), and the maximum radius of the m-th array

$$\epsilon_m := \max_{k=1\cdots n} \text{ rad } W_k^{(m)},$$

tends to zero such that

$$\epsilon_{m+1} \leq \frac{3}{\rho \eta} \epsilon_m^3 \qquad (m = 0,1,2,...)$$

where $\eta = \rho/(n-1)$.

Convergence will be further if one computes the auxiliary disks $V_k^{(m)}$ as soon as they are available. This amounts to replacing $V_k^{(m)}$ with

$$V_k^{(m)'} := \sum_{j=1}^{k-1} (z_k^{(m)} - W_j^{(m+1)})^{-1} + \sum_{j=k+1}^{n} (z_k^{(m)} - W_j^{(m)})^{-1} .$$

All these calculations are done by means of <u>circular arithmetic</u> according to Gargantini/Henrici (1971).

2. Refinement of the Schur/Cohn algorithm

2.1 Removing the roots of value 0

First, it must be mentioned that roots of value 0 have to be removed before applying the Schur/Cohn algorithm. 0 is a k-fold root of p, if and only if the k lowest coefficients $p_0,\ldots,p_{k-1}$ are equal to 0. This k-fold root is removed by dividing algebraically the polynomial by the factor z^k. Thus, in the following we may assume that the lowest coefficient $p_0 \neq 0$.

2.2 Scaling a polynomial

Now we look at the magnitude of the coefficients $\tilde{p}_j$ of a Schur transformation. Such a coefficient may be represented as a sum of 2 complex products each component being a sum of 4 real products. Those sums of products can be evaluated with maximal accuracy, since in PASCAL-SC a so-called long accumulator is provided for this purpose. Suppose that all the complex factors have (nearly) equal modulis m, the modulus of the sum maximally can be

$$M = 2m^2 ,$$

so that each component of the sum is absolutely $\leq M$. The iterated Schur transformations tend to produce overflow if $m > 0.5$. The modulus is limited by $M \leq m$ only if $m \leq 0.5$. This would prevent the use of the values > 0.5 of the machine number system. This can be improved by scaling.

Trivially, the roots of a polynomial remain unchanged if the polynomial as a whole is scaled by a factor f>0 being a power of 10. Likewise, the signs of the special coefficients that serve as test numbers remain unchanged by scaling. Thus, we have the opportunity of scaling.

A practical way is to scale the polynomial r (1.b) and every other Schur transform s of it in the following manner. The scaled polynomial s^* of degree m has the coefficients

$$s_i^* = f \cdot s_i \qquad (i=0,\ldots,m),$$

where f is the greatest machine power of 10 so that $f|s_i|<10^{24}$ $(i=0,\ldots,m)$. Then the moduli of the coefficients of the second Schur transform of s^* will be of a magnitude

$$M < 2 \cdot (2 \cdot (10^{24})^2)^2 = 8 \cdot 10^{96},$$

so that overflow is strictly avoided. On the other hand, the coefficients are made as great as possible under the circumstances given so that underflow is avoided as far as possible.

2.3 Maximum and minimum number of roots

What can be said about the location of polynomial roots if the result of the Schur/Cohn algorithm is "uncertain" (if some test number equals zero)? Following Henrici (1974), p.495, let h(q) denote the number of roots of a polynomial q inside the unit circle $|z|=1$, and $p^{(k)}$ the k-th Schur transform of the given polynomial p of degree n. Then there is an expression for $h(p^{(k+1)})$ in terms of $h(p^{(k)})$:

$$\text{If } \gamma_{k+1} \begin{bmatrix} > 0 \\ < 0 \end{bmatrix}, \text{ then } h(p^{(k+1)}) = \begin{bmatrix} h(p^{(k)}); \\ n-k-h(p^{(k)}). \end{bmatrix}$$

Reading these relations backwards yields

$$h(p^{(k-1)}) = \begin{bmatrix} h(p^{(k)}) & , & \text{if } \gamma_k > 0 \ ; \\ n+1-k-h(p^{(k)}), & & \text{if } \gamma_k < 0 \ ; \end{bmatrix}$$

These relations are used for the proof of theorem 1; they can also be used for deriving <u>a minimum and a maximum number of roots</u> of p in cases of uncertainty.

Let γ_{k+1} be the first test number γ_k with $\gamma_k = 0$. Then the K-th Schur transform $p^{(k)}$ is of a degree $\leq n-K$, and therefore, $h(p^{(k)}) \in \{0,1,\ldots,n-K\}$. Thus we can define

$$\text{hmin } (p^{(k)}) := 0 \qquad \text{and}$$
$$\text{hmax } (p^{(k)}) := n-K.$$

Accordingly, we define $\text{hmin}_k(p)$ and $\text{hmax}_k(p)$ to be the minimal and the maximal values of $h(p)$ on condition that γ_{k+1} is the first test number γ_i with $\gamma_i = 0$. As an auxiliary boolean function, we define

$$\text{neg}_k(p) := (-1)^{j-1} \text{ if } k_{j-1} \leq k < k_j \ (j=1,\ldots,m+1).$$

With respect to this definition, we supplement the definition of k_j by $k_0 := 0$ and $k_{m+1} := n+1$. We then find the following recursive relations:

<u>Lemma 1</u> (of Theorem 1).

$$\text{hmin}_0 (p) := 0 \ ; \qquad\qquad \text{and for } k>0:$$

$$\text{hmin}_k(p) := \begin{bmatrix} \text{hmin}_{k-1} (p) \\ \text{hmin}_{k-1} (p)+1 \end{bmatrix} \quad \text{if } \text{neg}_k(p) = \begin{bmatrix} -1; \\ +1. \end{bmatrix}$$

$$\text{hmax}_0 (p) := n \ ; \qquad\qquad \text{and for } k>0:$$

$$\text{hmax}_k(p) := \begin{bmatrix} \text{hmax}_{k-1} (p)-1 \ , \\ \text{hmax}_{k-1} (p), \end{bmatrix} \quad \text{if } \text{neg}_k(p) = \begin{bmatrix} -1; \\ +1. \end{bmatrix}$$

Proof by induction on k. The application of this lemma is demonstrated in the following example.

Example 2. (The values of γ_k are the same as in example 1.).Let n=9 and

γ_k =	9.3,	-3.1,	5.0,	2.7	-5.2,	1.3,	-1.7,	4.8,	-2.5		
for	k = 0,	1,	2,	3,	4,	5,	6,	7,	8,	9,	10.
	=	=			=		=		=	=	=
Then	k_0,	k_1,			k_2,		k_3,		k_4,	k_5;	
$neg_k(p)$ =	-1,	-1,	+1,	+1,	+1,	-1,	-1,	+1,	+1,	-1;	
$hmin_k(p)$ =	0,	0,	1,	2,	3,	3,	3,	4,	5,	5;	
$hmax_k(p)$ =	9,	8,	8,	8,	8,	7,	6,	6,	6,	5.	

If, for example, γ_8 is the first of the test numbers γ_k that equals 0, then $hmin_7(p) = 4$, $hmax_7(p) = 6$, and

$$4 \leq h(p) \leq 6.$$

Such an information about minimum and maximum number of roots will help to analyse the result of the bisecting process, especially in cases for which there are multiple roots or clusters of roots.

2.4 Interval arithmetic

For a strict verification by means of computers with rounding errors, the test numbers have to be calculated by interval arithmetic. This is, in fact, done by circular arithmetic since this is needed anyway for the Newton method (1.5). Thus, instead of real test numbers γ_k we will get real intervals Γ_k for the test numbers. The result of the Schur/Cohn algorithm will be "uncertain" as soon as some test number interval contains 0. Therefore, due to rounding errors, uncertainty will happen not only if some root is lying exactly on the unit circle but also if it is lying in an annular zone containing the unit circle.

For cases of uncertainty for some $\Gamma_{k+1} \ni 0$, the Schur/Cohn algorithm is broken off with a partial sum

$$h^* = h^*(p) = \sum_{j=1}^{\mu} (-1)^{j-1} (n+1-k_j)$$

where $k = k_\mu$ is the last of the calculated k_j with $\Gamma_{k_j} \not\ni 0$. When this occurs, my program displays the values $hmin(p) := hmin_k(p)$ and $hmax(p) := hmax_k(p)$ and stores them for a final analysis.

3. Refined bisecting process

3.1 Diamonds instead of squares

A machine representable square with sides parallel to the axes of the coordinate system would have an irrational semi-diagonal (= radius of the minimal disk covering the square). Rounding outwards those squares would produce a fourfold overlapping in the surroundings of the corners of the squares; this would be detrimental to the aim of bisection. I use diamonds, therefore, instead of the squares (see fig. 1). Subdividing a diamond into four subdiamonds will yield machine representable radii and coordinates of the midpoints provided that the radius of the original diamond (containing all roots of the polynomial) is chosen to have a mantissa being often enough divisible by 2. (For simplicity I will continue speaking of squares meaning diamonds).

With respect to the decimal floating point system of the SAM 68K with 13 place mantissa my program offers the opportunity to choose one of the following radius model mantissae $(0.1<a<1)$:

$$a = \begin{cases} 2^{20}/10^7, & \text{(default value)}, \\ 2^{30}/10^{10}, & \text{(option T)}, \\ 2^{40}/10^{13}, & \text{(option T T)}. \end{cases}$$

The program then chooses the minimum ρ of the values a, 2a, 4a, 8a and 10a, scaled by a suitable power of the base 10 so that $\rho > \sigma_0 \sqrt{2}$, σ_0 being the radius of a disk containing all roots of p. Thus, all roots of p are contained in the initial diamond having semi-diagonal ρ.

3.2 The test phase

Each bisecting step will produce a configuration consisting of subsquares. Each of these subsquares is included in the minimal disk covering the subsquare, and the Schur test is applied to the disk.

Since the semi-diagonal of all subsquares is the same and is stored once, only the midpoints of all subsquares are stored in an array D as long as the covering disks contain roots or are suspected to contain roots (1.3).

The sequence of midpoints in this array is important for a quick management. They are arranged in sloping lines running from north-west to south-east (see fig. 2). The sloping lines succeed each other from north-east to south-west, thus being comparable to the lines on a page.

When bisecting a configuration, first the northern and the eastern subsquares of all squares out of a given sloping line are tested; these subsquares may form a line of the new configuration. Hereafter the western and the southern subsquares of all squares out of the given sloping line are tested; they may form the subsequent line of the new configuration.

3.3 The separating phase

In order to separate the simple roots, the full algorithm by Schur/Cohn is needed. Time is spared if it is applied not only after the final bisection, but earlier. A special array F will store the separated inclusions of simple roots. There are 3 cases of separation of a simple root.

Case 1:   Simple root in a pillow region (see fig. 3a).

Let the minimal disk covering a square contain exactly 1 (simple) root, and all four neighbouring squares be proven to be free of roots. Then the root found is indeed contained in the middle square (and not in the outer parts of the covering disk), strictly speaking in a region similar in appearance to a pillow. Then the midpoint of the middle square is transferred from D to the array F and is cancelled in D.

Case 2:   Simple root in a lens region (see fig. 3b).

Let the minimal disks covering two neighbouring squares be proven or suspected each to contain 1 (simple) root, and let all six neighbouring squares be free of roots. We embed both these disks together in a greater disk whose midpoint is the middle of the midpoints of the given disks. Its radius is $> (1+\frac{1}{2}\sqrt{2})$ times the actual radius of inclusions.

If the greater disk contains exactly 1 (simple) root then the root is suspected to be common to both the smaller disks and to be situated in the overlapping region being similar to a lens. This suspicion is proven if the concentric middle disk having the same radius as the given disks likewise contains exactly 1 (simple) root. When this is the case, the midpoint of the middle disk is stored in F, and the midpoints of both the given disks are cancelled in D.

Case 3.:   Clover constellation (see fig. 3c).

Suppose that at least 3 of 4 squares having a common corner are covered by minimal disks that are suspected to contain a root (case of uncertainty), and none of the 4 squares be proven to contain a root. I call this a clover constellation. There is suspicion that the uncertainty is caused by a root lying in the near surroundings of the common corner of the 4 squares. Moreover, let all 8 squares neighbouring to the 4 given squares be free of roots. In order to test the suspicion, the 4 given squares altogether are embedded in a disk with double radius having as midpoint the common corner of the 4 squares. If the greater disk contains exactly 1 (simple) root the concentric disk with half the radius is tested. If this likewise contains exactly 1 (simple) root its midpoint is stored in F, and the midpoints of the 4 given squares are cancelled in D.

In the complex plane each pillow region (case 1) corresponds to two lens regions (case 2). The area of 1 pillow is to the area of 2 lenses approximately as 3 to 4. Thus, case 2 is more frequent than case 1. Case 3 normally is very rare; but its frequency will increase with the relative errors caused by rounding.

3.4 Neighbourhood relations (see fig. 4)

In separating the simple zeros many neighbourhood relations between squares have to be tested. To do this geometrically would cost a lot of computing time. Instead, book-keeping of neighbourhood relations is done from the beginning on by means of arrays like MU (mother square), TN (northern daughter square = subsquare), NW (north-western neighbour square). In these arrays each square is represented by the index it has in the D array. A square free of zeros is represented by the index 0 being not used in the D array. For example, let e be (the index of) the northern daughter of a square (with index) d. How do we find the north-western neighbour of e? This is the eastern daughter of the north-western neighbour of the mother d of e, as far as all participants are existing in the D array; otherwise, the wanted neighbour is free of zeros and thence is represented by the index 0.

3.5 When to switch to the separating phase?

Bisecting the squares whose midpoints are stored in the F array is more simple than those in the D array because neighbourhood relations are no longer needed in F. Sometimes only 2 or 3 subsquares have to be tested (instead of always 4 in the D array) since it is impossible that one root is contained in two opposite subsquares. Thus indeed, time is spared (as proposed in 3.3) if the separating phase is started early. On the other hand, it seems rather useless to start separating when only 2 bisecting steps (producing only 4·4 squares) are made. Thus, 3 bisecting steps at minimum are awaited until separation is begun.

3.6 When to stop bisecting?

In the F array, bisection of a given inclusion is stopped if no inclusion with half the actual radius of the corresponding root with multiplicity 1 is found. Here, not only the four subsquares are tested; if two or less of these have a certain multiplicity, the cases of a lens region or of a clover constellation are tested, respectively (3.3). If this fails too then the given inclusion is fixed and is transferred to the W array of resulting inclusions.

In the D array, the bisection process is at best stopped if all roots are simple and separated so that their inclusions are transferred to the F array. What, however, if this is impossible because of multiple roots or a cluster of roots?

When the inclusion radius becomes smaller and smaller, the relative errors by rounding tend to become greater and greater. Then, more and more test number intervals contain 0, and more and more subsquares are produced whose number of roots is uncertain. To put it in other words, the annular zone of uncertainty containing the borderline of a disk that includes a root will become broader and broader so that finally the whole disk is contained in a greater disk of uncertainty (see fig. 5).

Finally, it is impossible to verify by Schur/Cohn this root to be included in a disk having the actual inclusion radius. I call this state "inundation of uncertainty". Indeed, we find that from a certain step on, the number of "uncertain" subsquares increases exponentially.

What is the maximal number of "uncertain" subsquares that can occur under "reasonable" circumstances (without inundation of uncertainty)? If s (simple) roots are separated, then $n-s$ roots remain in the D-configuration. Let m be the sum of multiplicities of roots in the D-configuration. Each of these roots can contribute to 4 (certain or uncertain) inclusions maximally in the D-configuration; thus $m \leq 4 \cdot (n-s)$ as long as no inundation of undertainty has occurred. The bisection process in D should be stopped, however, if

$$m > 4 \cdot (n-s) \qquad \text{(stop formula)}.$$

3.7 Prolonged duration of the test phase

Sometimes one finds a great number of "uncertain" inclusions in the beginning of the bisecting process. This seems to occur mainly in squares neighbouring a cluster of roots, such as an inclusion of multiplicity ≥ 2. The uncertainty of the neighbouring inclusions will disappear again as soon as the roots forming the cluster are separated by further bisections.

Therefore, during the test phase the stop formula (3.6) is not applied. Additionally, the test phase is prolonged for a number of bisection steps in order to overcome this harmless sort of inundation of uncertainty. The total number of bisection steps during the testphase is called WAIT and is set experimentally to

$$\text{WAIT} = \max(3, g), \quad \text{where } g = {}^2\log(|p_n|/|p_0|),$$
ρ the initial radius of inclusions.

Experiments have shown that a number greater than 3 in this formula will lead to fruitless calculations.

3.8 Display of results.

Displayed results consist of several circles each followed by the number of roots enclosed in it. The results are trivial when all roots are simple and have been separated.

Otherwise, when the bisection process has been stopped by the stop formula (3.6), the final configuration will consist for the most part of several regions of coherent "uncertain" inclusions. Each of these regions will be enclosed in a suitable circle to which the Schur/Cohn algorithm is applied.

By those circles, roots are sometimes separated that could not ealier be separated while having inclusions with certain multiplicities!

For multiplicities >1, the displayed result needs a careful interpretation by the user because the following difficulties may occur:
1. A multiplicity may be uncertain. The minimal and the maximal number of roots are then shown.
2. A circle may happen to include or to overlap with another circle such that its multiplicity shown is greater than the number of roots in the area of coherent inclusions that has given rise to the enclosing circle. Such entanglements of circles will be displayed; the conclusions, however, are left to the user.

For example, if 10 is the degree of the polynomial, and there are 5 non-overlapping circles each enclosing at least 2 and at most 3 roots of the polynomial, then it is clear that each circle will contain exactly 2 roots.

Now the method may be summarized in the following

4. Solving algorithm

1. The user may enter a polynomial by its coefficients being complex numbers or circular intervals (options E, I) or complex random numbers (option R), or by its roots in order to test the program (option C); this last option will generally produce interval coefficients because of rounding errors. If the radii are set to 0 the roots may be changed more or less (1.1)!

2. The user may experimentally scale the roots of the polynomial (options / and *). He may choose a suitable radius model mantissa (3.1; option T).

3. Root finding is called by option Z.
   (For other possible options see menu!)

4. The polynomial p(z) is algebraically divided by z until the constant part becomes unequal to 0. The number of those divisions is stored as the multiplicity of the polynomial root 0.

5. Bisection of the complex plane as described.

5.1. Test phase, working with the D array only.

5.2. Separation phase, separated inclusions are transferred to the F array and further bisected there. Fixed inclusions are transferred from the F to the W array.

6. Refinement of results:

6.1. If all roots are simple and separated, and if the respective convergence condition is fulfilled, apply the cubically convergent simultaneous Newton procedure (option N). Otherwise, the bisection process is stopped by reaching a user defined accuracy or by the stop formula.

6.2. Display of the resulting inclusions (option S). If necessary, the roots are automatically re-scaled. Additionally, the medium and the maximum relative error of these inclusions are shown, as a measure for the accuracy.

7. When the results of a program run are not satisfying, it is optional to start a new run and
   - change the initial scaling of roots, or
   - change the grid of including squares (the radius model mantissa). This will be necessary in the very rare case that all the polynomial roots are lying on the corners of the tested squares so that the uncertainty is total.

5. Performance. Example.

The program is designed for polynomial degrees up to 20. Generally, random polynomials have simple roots only. These are solved in most cases if the degree is ≤15. The available accuracy is heavily diminished by clusters, because of inundation of uncertainty.

The time needed to separate all (simple) roots of a polynomial of degree N and to include them with maximal accuracy is about $3.5 \cdot N^3$ seconds. (Due to a special floating point arithmetic unit called BAP, this time is reduced to about $0.5 \cdot N^3$ seconds.)

For example, a polynomial was constructed by option C from its 16 simple roots without rounding errors and subsequently solved giving the following results. (The four place figures are rounded figures, for brevity.)

No.	Exact root	Found interval (center,	radius) for root
1.	(0, 10)	(8.937E-15, 10)	1.237E-14;
2.	(4, 0)	(4, 4.109E-15)	6.949E-15;
3.	(1, 1)	(1, 1)	0;
4.	(0, 1)	(3.906E-16, 1)	4.857E-16;
5.	(1, 0)	(1, 1.244E-14)	1.881E-14;
6.	(0.5, 0)	(0.5, -1.372E-17)	2.742E-17;
7.	(-1, 1)	(-1, 1)	0;
8.	(1, -1)	(1, -1)	0;
9.	(-1, 0)	(-1.000, 1.768E-15)	1.076E-13;
10.	(0, -1)	(2.125E-14, -1)	2.516E-14;
11.	(-1, -1)	(-1, -1)	0;
12.	(0, 0.25)	(1.201E-14, 0.25)	2.197E-14;
13.	(0, -0.2)	(4.244E-15, -0.2)	6.697E-15;
14.	(0, 1E-3)	(4.272E-94, 1E-3)	8.603E-93;
15.	(0.1, 0)	(0.1, -5.120E-17)	7.734E-17;
16.	(-0.01, 0)	(-0.01, 6.498E-18)	7.181E-18.

6. Conclusions.

The aim of my program has been to include the polynomial roots with strict verification. The computing time needed is relatively large; but no other method is known that can do this in such a global manner and with linear convergence. One purpose of this program may be to check the results of other, faster programs that will of need be developed in the future.

Possible improvements could be:
- Double length calculations, where necessary; the amount of fourfold time is diminished by avoidance of inundation of uncertainty.
- Separation of clusters of multiplicity ≥ 2 by special algorithms given by Gargantini/Henrici (1971) and Henrici (1974), p. 477-485.
- An implementation of the method suitable for parallel computers.

Literature

Bohlender, G. / L.B. Rall / Ch. Ullrich / J. Wolff v. Gudenberg (1986). PASCAL-SC. Wirkungsvoll programmieren, kontrolliert rechnen. Mannheim.

Brent, R.R. (1976). Multiple-Precision Zero-Finding Methods and the Complexity of Elementary Function Evaluation. Pages 151-176 in:

Traub, J.F. (1976). Analytic Computational Complexity. Academic Press. New York.

Collins, G.E. (1967). Subresultants and reduced polynomial remainder sequences. J.ACM 14, 1, 128-142.

Dejon, B. and P. Henrici (1969). Constructive Aspects of the Fundamental Theorem of Algebra. Wiley. New York.

Gargantini, I. and Henrici, P. (1971). Circular Arithmetic and the Determination of Polynomial Zeros. Numerische Mathematik, Band 18, S. 305-320.

Henrici, P. and Gargantini, I. (1969). Uniformly Convergent Algorithms for the Simultaneous Approximation of All Zeros of a Polynomial. Pages 77-113 in Dejon/Henrici (1969).

Henrici, P. (1974). Applied and Computational Complex Analysis. Volume I. Wiley. New York.

Kulisch, U. and Miranker, W.L. (1981). <u>Computer Arithmetic in Theory and Practice</u>. Academic Press. New York.

Kulisch, U. and Miranker, W.L. (Eds) (1983). <u>A New Approach to Scientific Computation</u>. Academic Press. New York.

Lehmer, D.H. (1961). A machine method for solving polynomial equations. <u>J.ACM</u> 8, 151-162.

Stewart, G.W., III (1968). <u>Some Topics in Numerical Analysis</u>. Oak Ridge National Laboratory. Oak Ridge, Tennessee.

Wilkinson, J.H. (1963). Rounding Errors in Algebraic Processes. Her Majesty's Stationary Office, London.

-    (1959). The evaluation of zeros of ill-conditioned polynomials. Part I. Numerische Mathematik 1, 150-166.

Verified Results for Linear Systems with Sparse Matrices

DIAMOND deliverable D3-3

W. Klein
Universität Karlsruhe

Abstract:
Self-validating computation for linear systems of equations with sparse matrices is presented with emphasis on matrices with a general sparse and a band-shaped structure. By introducing new data structures for the computer representation of these sparse matrices and by the use of PASCAL-SC, which supports an optimal computer arithmetic as introduced by Kulisch and Miranker, an algorithm for the determination of a verified enclosure of the solution of a linear system of equations is implemented. Existence and uniqueness of the true solution within the determined enclosure are automatically proved by the algorithm.

1. Introduction

The development and mathematical foundation of an exact computer arithmetic by Kulisch and Miranker [1] as well as the implementation of such a computer arithmetic and the embedding into the programming language PASCAL-SC [2] are the basis for new kinds of algorithms which produce verified results on a computer. Not only

(1) the determination of an interval ENCLOSURE of high accuracy of the true solution of a mathematical problem, but also

(2) the proof of EXISTENCE of the true solution within the calculated enclosure as well as

(3) the statement of UNIQUENESS with respect to the solution within these calculated bounds

are automatically guaranteed by these algorithms. The term "high accuracy"

means, that the infimum and the supremum of the component-wise computed
interval enclosure of the solution, containing the corresponding component
of the true solution, may differ only in the last bit of the mantissa of
the underlying floating point system. In order to obey these aims

> the mathematical standard functions,
> the evaluation of arithmetic expressions,
> the evaluation of polynomials and the determination of roots
>   of polynomials,
> the determination of eigenvalues and eigenvectors, as well as
> the solution of linear and nonlinear systems of equations

require a redesign and have to be implemented again.

For the implementation of algorithms which solve linear systems of
equations on digital computers, some more aspects must be considered:

- due to memory restrictions on computers an optimum representation of
  the problem data must be choosen. The representation of structured
  matrices with respect to the distribution of non-zero elements is
  especially important.

- in order to minimize the computational costs only the required
  arithmetic operations should be executed. The more information is known
  about the structure of the matrix, the better this reduction of
  arithmetic operations can be done.

Due to these aspects and the fact that in practise linear systems will
often appear in special structured form (e.g. the treatment of difference
methods and finite element methods) linear systems of equations may be
classified in the following way:

> linear systems of equations with a full non-symmetric matrix
> linear systems of equations with a full symmetric matrix
> linear systems of equations with a sparse matrix and
>   a symmetric pattern of non-zero matrix elements, or
>   a blockwise pattern of non-zero matrix elements, or

a general pattern of non-zero matrix elements, or
a band-shaped pattern of non-zero matrix elements, or
combinations of all these properties.

In this paper, we are concerned with linear systems of equations either with a general sparse or a band-shaped pattern of non-zero matrix elements.

We will consider a matrix to be <u>sparse</u> if it has sufficient zero elements so that it is worthwhile to use special techniques that avoid storing or operations on the zeros. Similarly, we will define a <u>band-shaped</u> <u>structure</u> of a matrix if all non-zero elements have a small distance from the main diagonal.
The following topics concerning sparse matrices are considered:
- an optimum algorithm with respect to the special matrix structure,
- a compact, directly accessible manner for storing the information,
- an optimum data structure for storing the non-zero elements of the general sparse matrix,
- exploitation of sparseness by eliminating unnecessary arithmetic operations,

When handling band-shaped matrices, aspects such as constant values in all diagonal elements or symmetry with respect to the main diagonal may be interesting. In particular for general sparse matrices, different ways for storing the non-zero elements such as (directed) graphs, pointers, which connect the non-zero elements to a linked list, arrays storing the non-zero elements as well as their corresponding row and column indices should be examined beforehand.

2. Method description

The problem of solving a linear system of equations $A * x = b$ with a real sparse nxn matrix A and a real point or interval vector b leads essentially to two different approaches:

a) determination of the inverse A^{-1} of the system matrix A followed by a multiplication of the inverse A^{-1} with the right-hand side vector b.

The determination of a matrix A^{-1} with

(4) $A^{-1} * A = E$ (E : identity matrix)

can be done, for example, with the Gaussian algorithm. If A is detected
to be 'invertible' (this should be confirmed by the feasibility of the
algorithm), then the solution x of the problem is computed by the
matrix vector product $(A^{-1} * b)$:

(5) $x = A^{-1} * b.$

b)   the determination of a LU-factorization of the system matrix A followed
     by a forward and backward substitution:
     The two matrizes $L = (l_{i,j})$ and $U = (u_{i,j})$ with the properties

     -   $L * U = A,$
     -   L is a lower triangular matrix,
     -   U is an upper triangular matrix, $u_{i,i} = 1$ for i= 1(1)n,

     are determined row-wise for i = 1(1)n according to the following
     formulas:

(6) $l_{i,k} = a_{i,k} - \sum_{j=1}^{k-1} (l_{i,j} * u_{j,k})$
         for $1 \leq k \leq i,$

(7) $u_{i,k} = \{ a_{i,k} - \sum_{j=1}^{i-1} (l_{i,j} * u_{j,k}) \} / l_{i,i}$
         for $i < k \leq n.$

The braced expressions in (6) and (7) can be computed with one exact
dot product requiring only one final rounding. Then, the solution x of
$L * U * x = b$ is computed by the forward substitution $L * y = b$

(8) $y_i = \{ b_i - \sum_{j=1}^{i-1} (l_{i,j} * y_j) \} / l_{i,i}$
         for i = 1, 2, ..., n

followed by the computation of the backward substitution $U * x = y$:

$$(9) \qquad x_i = y_i - \sum_{j=i+1}^{n} (u_{i,j} * x_j)$$
$$\text{for } i = n,\ n-1,\ n-2,\ \ldots,\ 1.$$

Again, the braced expressions in (8) and (9) can be computed with one exact dot product only.

In our special case of a sparse matrix, inversion usually results in a full matrix A^{-1}. This means, that all advantages of the sparse structure, for example the advantage of reduced storage requirement and reduced computation time by restriction to operations only between non-zero elements, are lost. Additionally, the implementation of the determination of the inverse matrix will require a new data structure beside the special constructions used for handling matrices either with a general sparse or a band-shaped pattern.

On the other hand, the LU-factorization of the system matrix A will preserve the sparse pattern of the matrix. This means in the case of a band-shaped matrix, that the matrices L and U of the LU-factorization have the same number of lower and upper co-diagonals as the band-shaped system matrix A. And similarly when handling general sparse matrices, that the LU-factorization will essentially preserve the general sparse structure provided the matrix has been suitably transformed such that only a few additional non-zero elements will appear in the lower triangular matrix L and the upper triangular matrix U. This effect is called "fill in".

Therefore, the actual LU-decomposition for general sparse matrices will be divided into two parts:

- the symbolic LU-factorization using only the indices of the non-zero elements which will determine the places, where the additional "fill in" in the matrices L or U might be possible and, therefore, will expand the data structure in these places, and
- the numerical LU-factorization in which the numerical values of A are used for the determination of the non-zero values of the elements of the lower triangular matrix L and the upper triangular matrix U.

Thus, the symbolic as well as the numerical LU-factorization does not require the introduction of an additional data structure, and an equivalent linear system of equations L * U * x = b results. This system can be solved easily by a forward and a backward substitution.

Therefore, the following self-validating algorithm for solving a linear system of equations A * x = b with a square sparse matrix A either with a general sparse or a band-shaped pattern is based on the LU-decompositon of the system matrix A.

3. Method implementation

In this implementation of the algorithm which computes an enclosure of high accuracy (see [1]), the programming language PASCAL-SC has been used. The following moduls are embedded:

$INCLUDE 'INTERVAL.PAK';
$INDLUDE 'GRAPHIK.PAK';

For the representation of a real interval with lower bound INF (infimum) and upper bound SUP (supremum) the following special data type

INTERVAL = RECORD INF, SUP : REAL END; ,

is introduced. For a variable X of type INTERVAL, these bounds are referenced by the notation X.INF or X.SUP. In order to support calculations with high accuracy, an exact dot product called SCALP is implemented in PASCAL-SC. This function SCALP computes the exact scalar product of two vectors X and Y without any rounding errors. In the case of a result which is not exactly representable as a floating point number, the determination of the dot product is followed by one rounding to the preceding, to the nearest, or to the succeeding floating point number in the computers floating point format (PASCAL-SC: 13 decimal digits in the mantissa and an exponent range from -99 to 99). Intermediate overflow and underflow conditions as well as the cancelation of leading digits of the mantissa are treated without loss of accuracy. The generally used notations for these roundings are ▽, □ and △ , for the rounding to the preceding, the nearest and the succeeding floating point number, respectively. The rounding of a number to the smallest enclosing interval with computer representable bounds is noted by ◇.

3.1 Implementation of the Algorithm

3.1.1 LU-Decomposition using intervals

If the data of the input matrix A and of the right-hand side vector b are point data, it is suitable to compute an interval LU-factorization $[L]$ and $[U]$ of the matrix A with

(10) $\qquad A \in [L] * [U]$.

{Here and in the following brackets, $[...]$, denote interval data}. According to the formulas (6) and (7), the interval enclosures $[l_{i,j}]$ and $[u_{i,j}]$ of the elements $l_{i,j}$ and $u_{i,j}$ of the exact LU-factorization are computed row-wise
for i= 1 (1) n:

(11) $\qquad [l_{i,j}] = ◇ \{a_{i,j} - \sum_{k=1}^{j-1} ([l_{i,k}] * [u_{k,j}])\}$

$\qquad\qquad$ for $1 \leq j \leq i$,

(12) $\qquad [u_{i,j}] = ◇ \{a_{i,j} - \sum_{k=1}^{i-1} ([l_{i,k}] * [u_{k,j}])\}$ / $[l_{i,i}]$

$\qquad\qquad$ for $i < j \leq n$.

Here, the optimum scalar product SCALP is used in connection with interval data (Function ISCALP): For the computation of the expressions

(13) $\qquad \{a_{i,j} - \sum_{k=1}^{j-1} ([l_{i,k}] * [u_{k,j}])\}$ $\qquad$ and

(14) $\qquad \{a_{i,j} - \sum_{k=1}^{i-1} ([l_{i,k}] * [u_{k,j}])\}$

with maximum accuracy in (11) and (12), respectively, the interval scalar product ISCALP is called for each term once, followed by the rounding $\diamondsuit$ to the smallest enclosing interval according to the computers floating point format.

In the case of the existence of the interval LU-factorization, such that no main diagonal element $[l_{i,i}]$ of the interval matrix $[L]$ contains a zero, the solution x of the linear system of equation $A * x = b$ is contained in the interval vector $[x]$, which is determined by forward and backward substitution

(15) $[L] * [y] = b$ and

(16) $[U] * [x] = [y]$

using interval arithmetic (see (8), (9)). The explicit formulas of (15) and (16) are

(17) $[y_i] := \diamondsuit \{b_i - \sum\limits_{j=1}^{i-1} ([l_{i,j}] * [y_j]) \} / [l_{i,i}]$

for $i = 1, 2, \ldots, n$, and

(18) $[x_i] := \diamondsuit \{[y_i] - \sum\limits_{j=i+1}^{n} ([u_{i,j}] * [x_j]) \}$

for $i = n, n-1, n-2, \ldots, 1$.

No division by main diagonal element $[u_{i,i}]$ is required, since they always have the value 1.0, due to the construction of the LU-factorization. In (17) and (18), the interval scalar product ISCALP is used for each of the expressions

(19) $\{b_i - \sum\limits_{j=1}^{i-1} ([l_{i,j}] * [y_j])\} / [l_{i,i}]$

and

(20) $[y_i] - \sum\limits_{j=i+1}^{n} ([u_{i,j}] * [x_j])$

in order to get a result of maximum accuracy, followed by one rounding $\diamondsuit$ to the smallest enclosing interval within the computer's floating point format.

An alternative and even superior solution to encode the sums in formulas (13), (14), (17) and (18) would be the use of dot product expressions as they are available in PASCAL-SC level 2.

The resultant interval vector $[x]$ with components $[x_i]$, $i = 1,\dots,n$, from (18) is a verified enclosure of the true solution. This is the most direct way to solve the linear system $A * x = b$ with a band-shaped matrix A.

Even though all intermediate steps are done with maximum accuracy, the final intervals will often have a large span. An improvement of the quality of the computed result is possible by the following process of an iterative residual correction:
Suppose $\tilde{x}_1$ is an approximation of the solution x in $A * x = b$, determined from the already calculated interval enclosure $X:=[x]$ in (18), for example by

(21) $\qquad \tilde{x}_1 = 0.5 * (X.INF + X.SUP)$.

Let $[r]$ denote the interval enclosure of the residue defined by

(22) $\qquad [r] = \diamondsuit (b - A * \tilde{x}_1)$,

using one call of the exact scalar-product.

Then the following linear system,

(23) $\qquad [L] * [U] * [x_2] = [r]$

yields an enclosure $[x_2]$ of the correction term which together with $\tilde{x}_1$ gives an enclosure $[x]$ of the solution x of $A * x = b$ of the following form:

(24) $\qquad x \in [x] := \tilde{x}_1 + [x_2]$.

The steps (21) - (24) are repeated as long as certain stopping criteria are not satisfied (for example, the width of [x] might be a stopping criteria).

Practical experience with this program shows, that the interval LU-decomposition should only be used in the case of a "well conditioned" system matrix A, for example, a matrix which is almost diagonally dominant, and if a point vector is given for the right-hand side.

3.1.2 Approximate LU-decomposition

In the cases of an interval right-hand side vector b or of an ill-conditioned problem matrix A, the method is based on an approximate LU-factorization of the system matrix A:

$$(25) \qquad A \approx L * U \qquad (see \ (6),(7)) \ .$$

If this decomposition exists, i.e. there is no main diagonal element $l_{i,i}$ in L equal to zero, then an approximate solution $\tilde{x}_0$ of the solution x of the system $A * x = b$ is determined by solving

$$(26) \qquad L * U * x = b \qquad (see \ (8),(9))$$

via forward and backward substitution. The resultant approximation $\tilde{x}_0$ may be improved by repeated constructions of corrections $\tilde{x}_i$, i=1,2,...,m determined from linear systems of equations

$$(27) \qquad L * U * x_i = r_i$$

envolving the exact evaluation of the residue

$$(28) \qquad r_i : = \Box \ (b - \sum_{k=0}^{i-1} A * \tilde{x}_k) \ .$$

Then, the final approximation $\tilde{x}$ is represented by a sum of partial corrections $\tilde{x}_i$,

(29) $\tilde{x} = \sum_{i=0}^{m} \tilde{x}_i$,

which, of course, is not evaluated in order to keep the information of more than one mantissa length. Depending on the input data, interval or real right-hand side data vector, an enclosing interval vector [r],

(30) $[r] = \diamond (b - A * x)$, or
(31) $[r] = \diamond ([b] - A * x)$,

is computed. With this interval valued residue, an iteration of the following form is started:

(32) $[x_0] := [r]$
(33) $[z_i] := (1 + eps / 2) * [x_i] - eps / 2 * [x_i]$
          {epsilon inflation}
(34) $L * U * [x_{i+1}] = [r] - [A - L * U] * [z_i]$
          $i = 0,1,2,...$

Steps (33), (34) are repeated as long as $[x_{i+1}]$ is not strictly contained in $[x_i]$ and other stopping criteria are not satisfied. If the inclusion condition

   $[x_{i+1}] \subset [x_i]$

is satisfied, then the solution x of $A \cdot x = b$ is contained in

   $[x] := (\sum_{k=0}^{m} \tilde{x}_k) + [x_{i+1}]$.

The convergence of the iteration in (33) and (34) essentially depends on the quality (that means the width of intervals) of the interval residue [r] and of the interval matrix [A - L * U].

Here again, dot product expressions or one single interval scalar product are applied.

The method described above is implemented on a SAM 68K computer by KWS. The source programs are written in PASCAL-SC and are named BAMS and GEMS, which stand for "BAnd-shaped Matrix System solver" and "GEneral sparse Matrix System solver". As mentioned earlier, both programs use the same algorithm with different datatypes for the representation of the non-zero elements of the system matrix. We, therefore, will not list the main procedures of these programs in this paper (in contrast to [5]) but the algorithm upon which both programs are built:

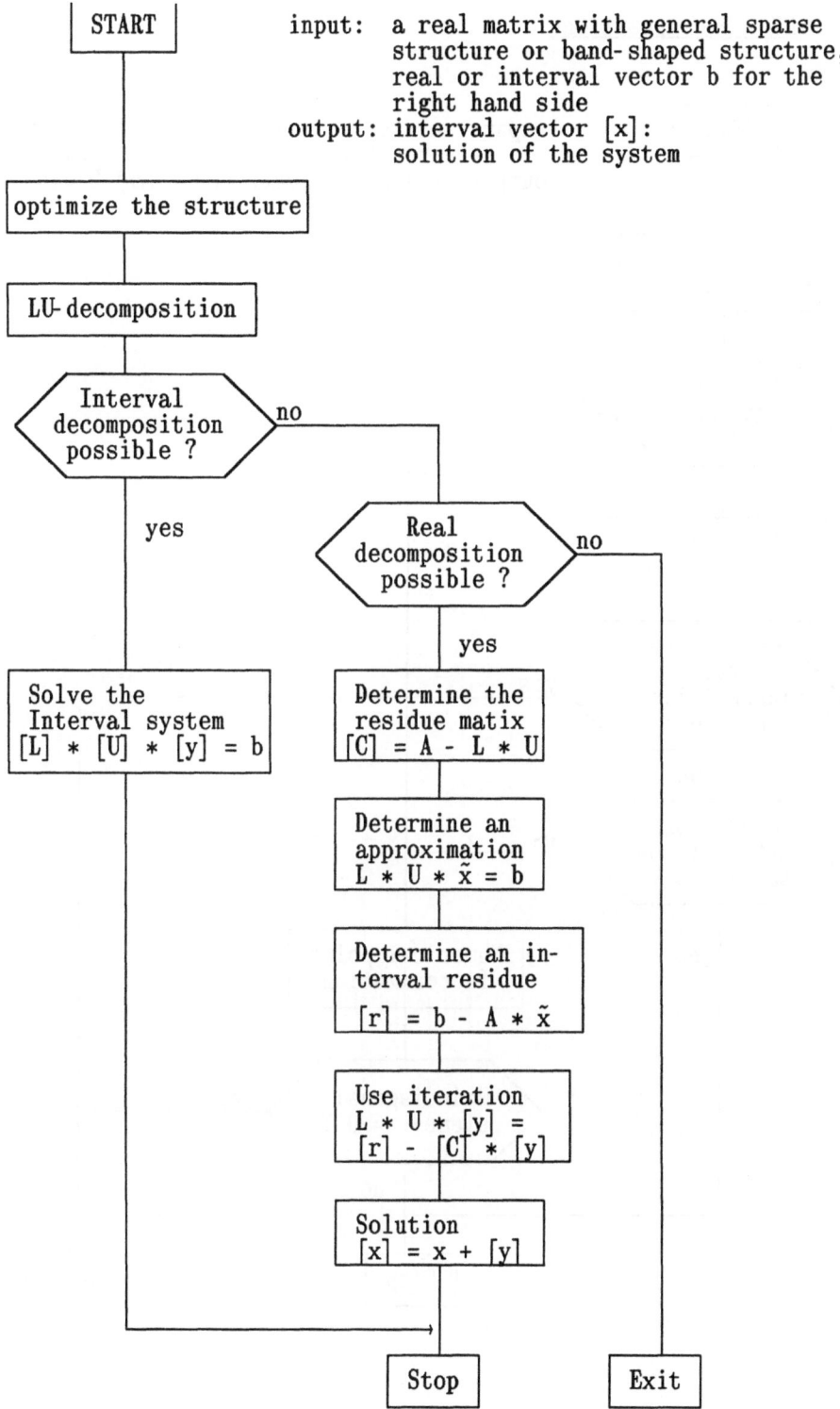

LU - Decomposition

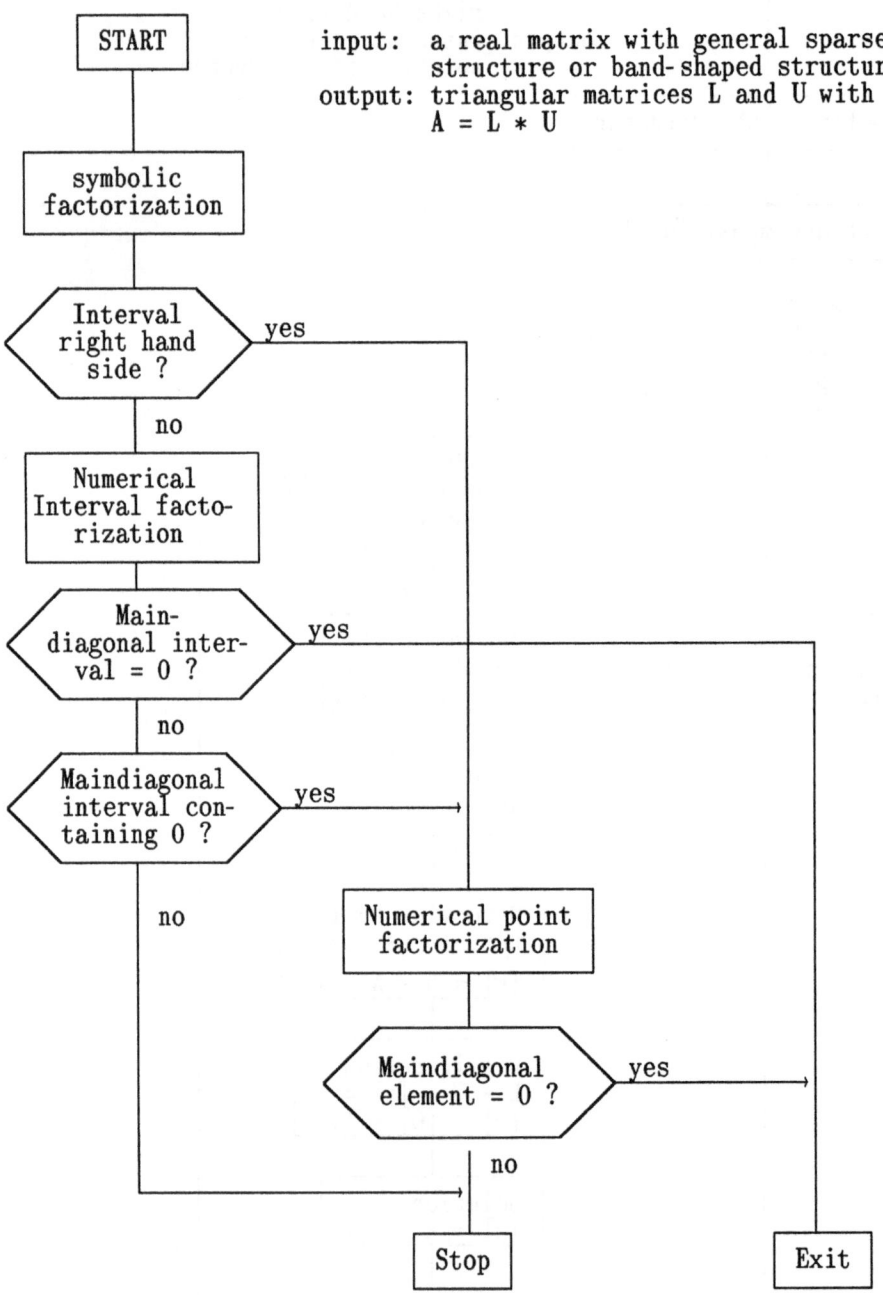

input: a real matrix with general sparse
structure or band-shaped structure
output: triangular matrices L and U with
A = L * U

Determine an Approximation

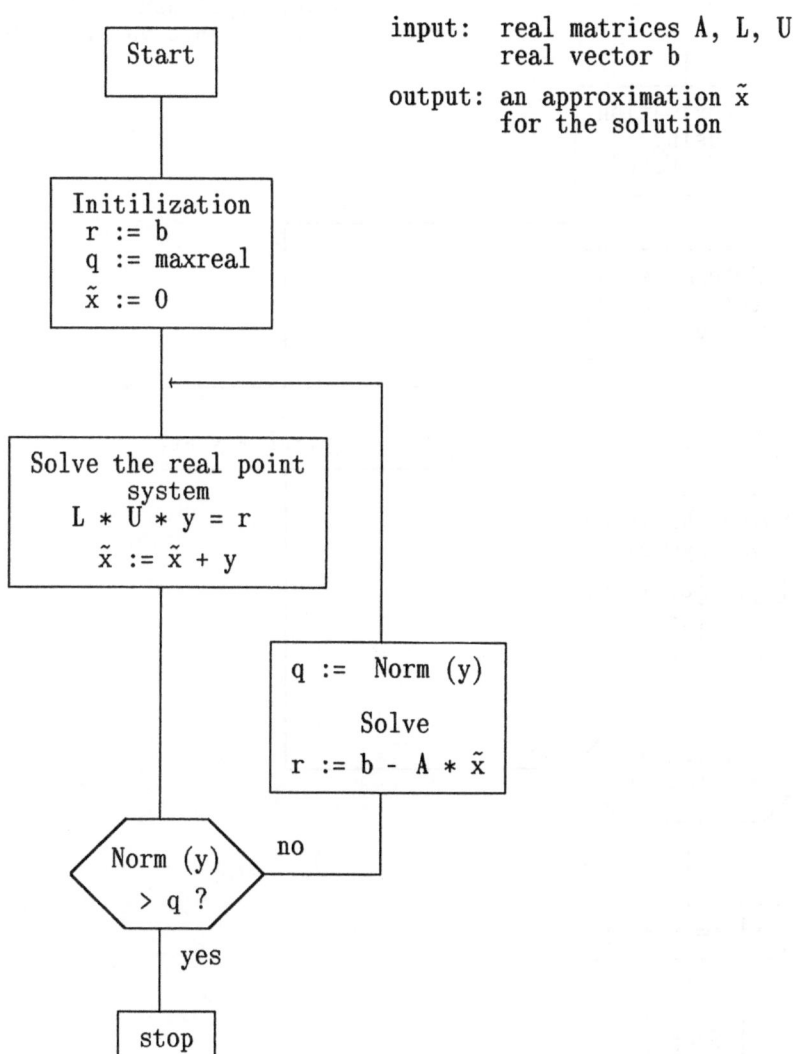

input: real matrices A, L, U
        real vector b

output: an approximation x̃
        for the solution

Solve an interval system [L] * [U] * [x] = b

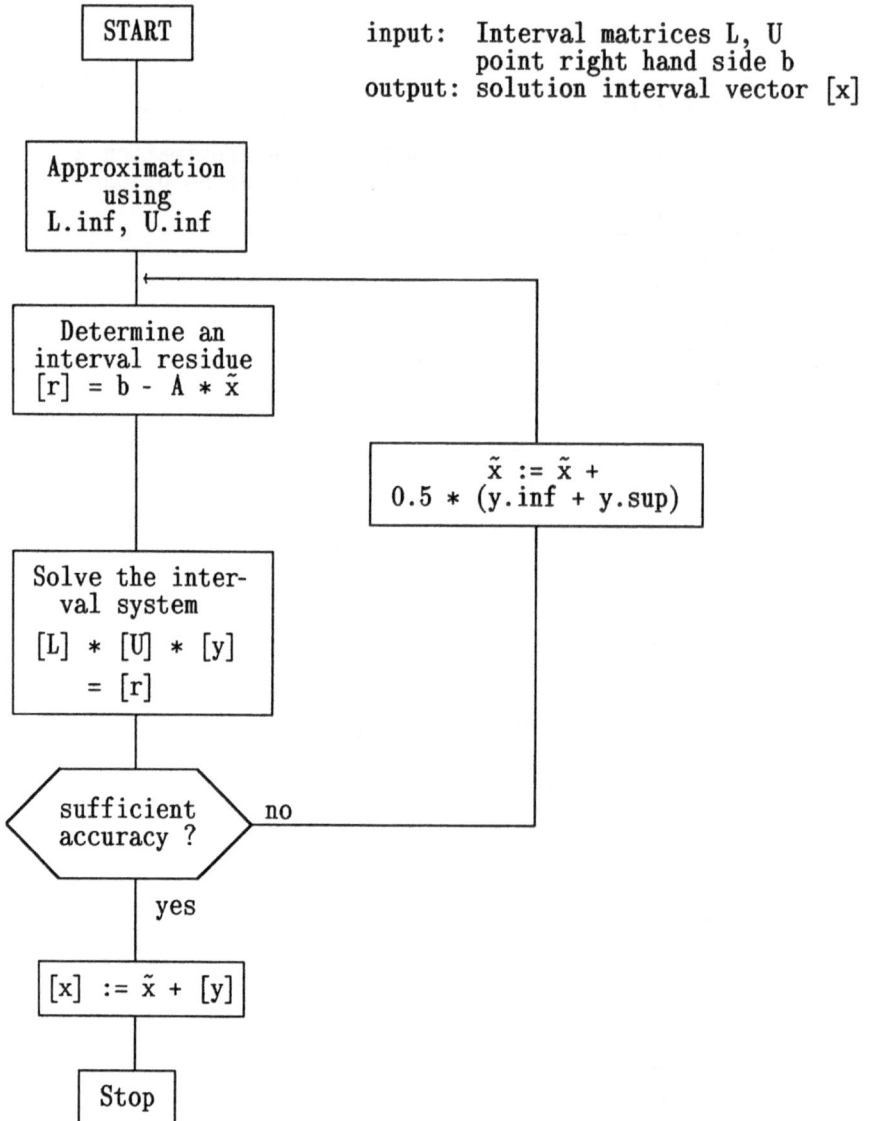

input: Interval matrices L, U
        point right hand side b
output: solution interval vector [x]

3.2 Data structures for general sparse matrices

As mentioned earlier, there are three different ways for storing the non-zero elements of a general sparse matrix.

(1) Directed graphs,
(2) pointers, which connect the non-zero elements to a linked list,
(3) and arrays storing the non-zero elements as well as their corresponding row and column indices

should be considered first: For the implementation of an algorithm which solves linear systems of equations with general sparse matrices, it is important to consider the way in which the elements $a_{i,j}$ of a sparse matrix A = $(a_{i,j})$ are addressed. Inserting or deleting non-zero elements in a data structure are standard operations concerning sparse matrices.

ad (1): With sparse matrices, it is often useful to associate matrices with graphs in such a way, that each row is represented by a node. Then, the non-zero element $a_{i,j}$ is represented by a connection from node i to node j. (Connections are also called edges of the graph). For symmetric matrices, undirected connections between the nodes i and j can be used. Matrices with a band-shaped pattern are easy to represent by graphs using a periodic data structure. Data structures using graphs are also useful to determine the block-lower-triangular form of a general sparse matrix A (algorithm of Tarjan or Sargent/Nesterberg).

ad (2): The simplest way of working with linked lists is to use a single list for all non-zero elements which connect the non-zero elements via pointers. It is very easy to insert a new or delete an unnecessary element by only adjusting the pointers. The disadvantage of this method is the expensive manner of addressing one special element $a_{i,j}$ of the matrix A. A better way is to use more than one linked list, one list for pointing to the first element in each row (or column) and one list for all elements in a row (or column).

ad (3):   The most convenient way to store the non-zero elements of a general sparse matrix is to keep them in form of a triple $(a_{i,j}, i, j)$ which combines the real value of an element with the corresponding row and column indices of the element. An implementation could be realized using one array of real numbers for the numerical values of the elements and two integer arrays for the row and column indices where the real and integer arrays have the same size. But as mentioned above, inserting or deleting an element will require time consuming shift operations of all arrays. The data structure used in this paper consists of one integer array IA with length equal to the dimension of the matrix + 1 for storing the position of the first elements of each row, and two arrays of length L_1

due to the number of non-zero elements: the integer array JA holds the column indices of the non-zero elements and the real array A the numerical values of the non-zero elements.

Example:

$$\begin{bmatrix} 2 & 0 & 3 & 0 & -1 & 0 \\ 2 & 8 & 0 & 0 & 0 & 0 \\ 0 & 0 & 1 & 0 & 0 & 0 \\ -3 & 0 & 0 & 5 & 0 & 0 \\ 0 & -1 & 0 & 0 & 4 & 0 \\ 0 & 0 & 0 & 0 & 2 & 8 \end{bmatrix}$$

This matrix will be stored in the three vectors

IA = (1 4 6 7 9 11 13)
JA = (1 3 5 1 2 3 1 4 2 5 5 6)
A = (2 3 -1 2 8 1 -3 5 -1 4 2 8)

In addition, there are many other possibilities for storing the non-zero elements in combining arrays and linked lists: one way is to store a pointer to the first elements of a row in an array whose elements consists of linked lists for the other elements in that row represented by a record consisting of the numerical value and the column index of an element of the matrix.

Benchmark tests on a SAM-68K computer for a matrix vector multiplication y = A * b with several data structures T1, T2, T3 for the matrix A brought the following results:
The tests T1 and T2 work with the data structure described ad (3) above.

For test T1: the non-zero elements of the i-th row of A, i = 1..n, are collected into one vector (A_i), the resulting element $y_i = \sum_{j=1}^{n} A_{ij} * b_j$ is determined by one call of the scalar product function scalp for the two vectors (A_i) and (b). Using test T2, the resulting element is determined by repeated calls of the accumulating scalar product function scalp. This method is the quickest way for the determination of a matrix-vector product. For test T3, we use an array [1..n] or linked lists to store the matrix A: Each non-zero element of one row is stored as a record of its value, column index and a pointer to the next non-zero element.

All tests were executed for three dimensions and three kinds of matrices: a sparse matrix (10% of the elements are non-zero), a matrix with less than 40% non-zero elements and a full matrix; the measure of time is milliseconds.

Test 1:   One call of the scalar product function with two vectors with length equal to the number of non-zero elements.

Test 2:   Repeated call of the accumulating scalar product function .

Test 3:   Use of linked lists.

Time in millisec.

dimension	number of non-zero elements	T1	T2	T3
dim = 3	elem = 3	236	38	101
dim = 3	elem = 5	238	41	103
dim = 3	elem = 9	241	45	107
dim = 10	elem = 10	707	49	108
dim = 10	elem = 39	739	92	150
dim = 10	elem = 100	782	146	205
dim = 50	elem = 50	3400	106	152
dim = 50	elem = 1000	4180	1137	1163
dim = 50	elem = 2500	5412	2789	2783

This example demonstrates that the accumulating scalar product function using two one dimensional vectors is the quickest way on the SAM 68K computer to determine the matrix vector product A * b.

3.3 Data structures for band-shaped matrices:

There are many possibilities for storing the non-zero elements of a matrix with a band-shaped pattern. We assume that the matrix is a DIMxDIM band matrix with -NU lower co-diagonals and NO upper co-diagonals. In this program, one dimensional vectors of type RBAND or IBAND are used for storing the point or interval values of the non-zero elements of the occuring matrices:

- the storage of each co-diagonal requires DIM locations of the vectors of type RBAND or IBAND

- the first lower co-diagonal with the elements $a_{i,j}$, $j-i = NU$, is stored at the first DIM locations of the band (DIM = actual dimension), then the second of the lower co-diagonals with elements $a_{i,j}$, $j-i = NU + 1$, at the next DIM locations, and so on, up to the last of the upper co-diagonals with elements $a_{i,j}$, $j-i = NO$.

- the i-th lower co-diagonal is stored with i leading zeros at the beginning of the reserved DIM locations, the i-th upper co-diagonal is stored with i trailing zeros at the end of the reserved DIM locations.

- a total of $(NO-NU+1) * DIM$ locations beginning with the index 1 of the vectors of type RBAND (IBAND) are used.

Example: the components of the 6x6 matrix $A = (A_{i,j})$

$$A = \begin{bmatrix} 2 & 3 & 4 & 0 & 0 & 0 \\ 1 & 2 & 3 & 4 & 0 & 0 \\ 0 & 1 & 2 & 3 & 4 & 0 \\ 0 & 0 & 1 & 2 & 3 & 4 \\ 0 & 0 & 0 & 1 & 2 & 3 \\ 0 & 0 & 0 & 0 & 1 & 2 \end{bmatrix} \quad ,$$

are stored in the following way in an array B of the type RBAND:

```
DIM = 6      { actual dimension }
NU  = -1     { number of lower co-diagonal, negative! }
NO  = 2      { number of upper co-diagonals }
```

```
index 1              7            13           18
B   = |0 1 1 1 1 1|2 2 2 2 2 2|3 3 3 3 3 0|4 4 4 4 0 0|

        first lower     main     first upper  sec. upper
        co-diagonal   diagonal            co-diagonals
```

The following relation holds between the indices i and j of an element $A_{i,j}$ of the given matrix A and the index k of the corresponding element B[k], holding the value of the element $A_{i,j}$:

(35) A[i , j] = B[k]
      k := i + (j - i - NU) * DIM

with $NU \leq i-j \leq NO$.

3.4 LU-Decomposition for general sparse matrices

As mentioned earlier, the handling of general sparse matrices leads to an LU-factorization, which is divided into two parts:
  - the symbolic LU-factorization using only the indices of the non-zero elements and
  - the numerical LU-factorization for the determination of the numerical values of the matrices L and U.

There is one important problem in the LU-factorization of a general sparse matrix: a non singular sparse matrix need not have a decomposition with the properties written above:

Example :

There is no LU-factorization for the matrix $A := \begin{bmatrix} 0 & 1 \\ 1 & 0 \end{bmatrix}$. Applying a permutation $P = \begin{bmatrix} 0 & 1 \\ 1 & 0 \end{bmatrix}$ first,

the matrix $\tilde{A} := P * A = \begin{bmatrix} 1 & 0 \\ 0 & 1 \end{bmatrix}$, will possess a LU-decomposition.

It can be shown constructively by the Gaussian algorithm that each quadratic non-singular matrix $\tilde{A} := P * A$, with P a suitable permutation matrix, possesses a LU-factorization of the required type. Therefore, in cases where the LU-decomposition failes not because A is singular, but because of the way in determining the matrices L and U (without using permutations), we have to look for suitable algorithm to determine a matrix P with the aim

- to enable a LU-factorization and, in addition,
- to reduce the "fill in",
- to reduce rounding errors,
- to reduce the number of required arithmetic operations, but
- to maintain numerical stability.

Such an algorithm is described in the paper of Gustavson [3], using a strategy of Markowiz.

3.5 Symbolic LU-factorization for general sparse matrices:

As mentioned above, we first have to determine the places where a new nonzero element may appear in the matrices L or U in comparison to the nonzero elements of the system matrix. Afterwards, we will enlarge the data structure of the matrix L or U at these places before we are able to compute the numerical values of these elements. Using the formula for the numerical LU-factorization,

$$l_{i,k} = \{ a_{i,k} - \sum_{j=1}^{k-1} (l_{i,j} * u_{j,k}) \}$$
$$\text{for } 1 \le k \le i, \qquad \text{and}$$

$$u_{i,k} = \{ a_{i,k} - \sum_{j=1}^{i-1} (l_{i,j} * u_{j,k}) \} / l_{i,i}$$

for $i < k \leq n$,

it is easy to see that in the case $a_{i,k} \neq 0$ the corresponding element $l_{i,k}$ in the matrix L or $u_{i,k}$ in U is zero only in the case of

$$a_{i,k} = \sum_{j=1}^{k-1} (l_{i,j} * u_{j,k}) \quad \text{or} \quad a_{i,k} = \sum_{j=1}^{i-1} (l_{i,j} * u_{j,k}).$$

The evaluation of elements $l_{i,j}$ or $u_{i,j}$ with zero value will not be treated, thus allowing certain elements of the LU-decomposition to be zero. Further on, new elements in the i-th row of the matrix L and U will occur whenever

$$\sum_{j=1}^{k-1} (l_{i,j} * u_{j,k}) \neq 0 \quad \text{for } k \leq i$$

and

$$\sum_{j=1}^{i-1} (l_{i,j} * u_{j,k}) \neq 0 , \text{ respectively.}$$

Due to the fact that the elements in the first row of U and in the first column of L may only change their values, but no additional "fill in" will appear in the first row and columm, the symbolic factorization can start at the second row of the sparse matrix A. For all elements $l_{i,j}$ of row i with $l_{i,j} \neq 0$ and $1 \leq j < i$ we have to consider the j-th row of L, if $l_{j,k}$ is an element with $l_{j,k} \neq 0$, $j < k \leq dim$ and $a_{i,k} = 0$. In this case the data structure has to be enlarged at this place. For a given n by n system matrix

$A = ((a_{i,j}))$, $1 \leq i, j \leq n$ we will describe now the

Algorithm :
Store all non-zero elements of A in the matrices L and U.

```
FOR i := 2 TO dim DO     { handle with each row }
   FOR j := 1 TO i-1 DO
      IF l_i,j ≠ 0 THEN      { consider row j }
         FOR k := 1 TO dim DO
            IF ( l_j,k ≠ 0 ) AND ( l_i,k = 0 ) THEN
               { enlarge the data structure at the place i, k }
```

Example :

For a given (6 by 6) matrix, the given non-zero elements are denoted by 'x' and the additionally appearing non-zero elements are denoted by '*'. Here, row 1 to 3 have been considered before.

$$\begin{bmatrix} x & . & x & . & x & . \\ x & x & * & . & * & . \\ . & . & x & . & . & . \\ x & . & * & x & * & . \\ . & x & * & . & x & . \\ . & . & . & x & * & x \end{bmatrix}$$

Row 4: The first non zero element appears in columm $j = 1$. Although the element at columm $k = 1$ in row $j = 1$ is non-zero, the property $k = j$ holds. Therefore, the next co-

lumns $k = 3$ and $k = 5$ in row $j = 1$ will be considered. Because of $l_{4,3} = 0$ and $l_{4,5} = 0$ it is likely that there will appear new non-zero elements in the matrices L and U; thus new places in the data structure have to be reserved.

4. Remarks:

- in future the programs may be extended in the following way: the components of the input matrix will be intervals

- both programs will be rewritten in the new PASCAL SC Version 2.0 for SAM 68K machines.

References:

[1] Kulisch, U/Miranker, W: Computer Arithmetic in Theory and Practice, Academic Press, New York (1981).

[2] Bohlender, G/Rall, L. B./Ullrich, Ch./Wolff v. Gudenberg, J.: PASCAL-SC, BI Wissenschaftsverlag (1986).

[3] Gustavson, F.: A Survey of some sparse Matrix Theory and Techniques, Jahrbuch Überblicke Mathematik 1981, S. 63-105.

[4] Gustavson/Lininger/Willoughby: Symbolic Generation of an optimal Crout Algorithm for sparse Systems of linear Equations, JACM 17 1970, pp.87-109.

[5] DIAMOND paper Doc. No.: D3-3/1 "Verified Results for Linear Systems with Band-shaped Matrices" (12/17/86).

Self-Validating Numerical Quadrature

DIAMOND Deliverable D3-5
Rainer Kelch
Universität Karlsruhe

Abstract: The goal of this paper is to create a numerical procedure which, for a large class of functions, validates numerically the definite integral

$$J := \int_a^b f(x) \ dx$$

in small bounds. In principle, there are three different methods for computing J. Let us choose the one computing an approximation Q for J by means of a numerical quadrature formula and enclosing the procedure error in the error term R such that:

$$J \in Q + R.$$

To calculate Q or R on the computer such that the relation mentioned above is valid, a corresponding interval arithmetic [1], [7] is necessary as well as its realization in an appropriate programming language [2]. To achieve the above goal, new developments and ideas were cultivated upon three essential points:

-   In general, to validate R numerically, the value of an arbitrarily high derivative of the integrand function f on an interval has to be computed via automatic differentiation algorithms [2], [3], [4], [9], [10], [11]. They are implemented for all standard operations and functions.
-   As a quadrature procedure, the Romberg-Extrapolation [12] was modified such that instead of the standard iterative computation of Q via the T-table only <u>one</u> direct evaluation of a scalar product becomes necessary. This implies not only shorter runtime, but also higher accuracy.
-   Furthermore, accuracy and fastness are improved via adaptive refinement to reduce the total error in the integration domain.

Irrespective of the requested accuracy, this procedure always provides an interval which guarantees the enclosure of J. The procedure was implemented in such a way that, in principle, each function chosen out of $C^\infty[a,b]$ was able to be integrated.

1. Review

For a large class of functions f ∈ $\mathscr{F}$ a numerical procedure is to be created verifying numerically the value J of the definite integral on f in small bounds. The interval boundaries a and b are to be representable on the computer. For $\mathscr{F}$ let us choose the set of functions $C^{\infty}[a,b]$ that are infinitely often continuous differentiable on [a,b]. As we see below, $\mathscr{F}$ may also be extended to C^{2m+2} [a,b]. Chapter 2 briefly deals with the necessary fundamentals. In this chapter, we survey the three principally different methods of computing J. We decided in favour of the method which allowed the inclusion of J by means of the sum of the quadrature-approximation Q and the inclusion of the procedure error R. Afterwards, we deal with a few facts of Computer-Arithmetic which have to be considered in a verified computation of J. Finally, the implementation of the Computer-Arithmetic in the programming language PASCAL-SC is presented.

Chapter 3, 4 and 5 describe the three essential items from which new ideas are cultivated. They require the largest programming size and programming effort. The mathematical fundamentals, algorithms and implementation of the differentiation package are described in chapter 3. Arbitrarily high derivatives of arbitrary functions of $\mathscr{F}$ are verified numerically. Their simple handling in programming is only made possible by the new operator-, function- and data-concept of PASCAL-SC.

Chapter 4 shows how the Romberg-Quadrature was modified in such a manner that instead of the standard iterative computation of Q via the T-table only one direct evaluation of a scalar product is necessary. For computer implementation the required data type specified in the differentiation package (see chapter 3) is also considered.

Chapter 5 deals with an improved "bisection" to achieve a reduction of the computing time and of the total error. This may be achieved both by the choice of a corresponding T-table element and by an adaptive refinement, under the condition that good reliable error bounds are available. So an effective error control is now possible.

In chapter 6, numerical results are discussed. In some cases the necessity of enclosure algorithms becomes obviously.

2.   Fundamentals

2.1 Different Quadrature Methods

In principal, there are three different methods for numerically validating a definite integral:

(1) **Formal Integration with Subsequent Evaluation**

Formal integration, i.e. the definition of the primitive function, is made possible via Computer Algebra systems. In general, however, there is no closed representable primitive function for each integrand. If a primitive function is computed it still has to be evaluated numerically.

(2) **Computation in the Functoid**

First the integrand function f must be semimorphly projected into the screen function made possible by the representation of f as a finite linear combination of basic elements of the corresponding functoids [4], [5], [6], [8]. Due to this finite representation, the procedure error is added to the screen function $IS_N f$ intervalwise. Afterwards $IS_N f$ is integrated formally in the functoid using the existing integral operator. Hence, the result is a screen-primitive, which subsequently has to be evaluated.

(3) **Numerical Approximation and Inclusion of the Procedure Error**

Note that a primitive function is not computed. A linear combination of function values at certain points, $x_i \in [a,b]$, and with certain weights, w_i, results in a numerical approximation Q. The procedure error is enclosed in the remainder term R, such that:

$$J \in Q + R .$$

In general, a higher derivative of the integrand function has to be evaluated. The definition of these derivative values constitutes the main problem for the verified computation of J. In view of the alternative between formal and automatic computation, we settled on automatic differentiation (see chapter 3).

As a matter of course, method (1) is not the appropriate choice for our problem. Compared with method (2), method (3) has the advantage that this kind of numerical integration, though without verification and computation of the procedure error, is already well-known in many variations. Computing in the function screen may deal with a wide range of numerical problems. Method (3), however, only considers numerical integration. In particular, method (3) is applied for the solution of our specific problem.

2.2 Computer-Arithmetic, Rounding Error and Interval-Computing

Statements and formulas of numerics cannot simply be transferred to computers, due to rounding errors resulting from the finiteness of the machine numbers. These problems are overcome by a temporary computer-arithmetic with directed roundings and interval computing. In [1], [4], [7], [8] fundamentals, notations and formalisms are considered in detail.

The following identity is a simple and obvious example:

(1) $(1/3) * 3 = 1$

If the left side of equation (1) is evaluated on the computer, in general, 1 is not the result.

This problem arises in cases of numerical quadrature:

If the definite integral

$$(2) \qquad J := \int_a^b f(x)\,dx$$

is expressed by

(3) $J = Q + R$

with the approximation Q according to

$$(4) \qquad Q := \sum_{i=1}^{N} w_i * f(x_i) \quad , \quad x_i \in [a,b]$$

and the error term R according to

(5) $R := K* f^{(j)}(\xi)$, $\xi \in [a,b]$,

then, if Q and R are computed via a standard floating-point arithmetic, the equation (3) no longer holds. Strictly speaking, R cannot be computed in the form defined above, since ξ is unknown. A reliable statement is only possible if the rounding errors are compensated for by replacing the ordinary floating-point operations by interval operations [7]. The procedure error R must be enclosed by

(5') $\hat{R} := K * f^{(j)} ([a,b])$.

and the operations here, of course, also must be replaced by the corresponding interval operations.

That is, the value of J is validated numerically in so far as in (3) R is replaced by $\hat{R}$ and then the operations o are replaced by $\diamondsuit$. Thus, it holds that

(6) $J \in Q + \hat{R} \subseteq \tilde{Q} \diamondsuit \tilde{\hat{R}}$

with

(7) $\tilde{Q} := \{ \diamondsuit \sum_{i=1}^{N} w_i \diamondsuit f(x_i) \supseteq \diamondsuit (\sum_{i=1}^{N} w_i * f(x_i)) \supseteq Q$

and

(8) $\tilde{\hat{R}} := K \diamondsuit \diamondsuit f^{(j)} ([a,b]) \supseteq \diamondsuit \hat{R}$.

2.3 Realization of the Computer-Arithmetic in PASCAL-SC

The solution of the problems dealt with in 2.2 is made possible by PASCAL-SC, a scientific extension of the programming language PASCAL. PASCAL-SC provides among other things the following new language elements:

- Arithmetic operations with controlled rounding,
- Optimal scalar product,
- Functions with arbitrary result type,
- Realization of abstract data type via operator concept,
- Overloading of procedures, functions and operators,
- Dynamic arrays,
- Modul concept.

Furthermore, predefined standard packages such as those for computing complex numbers, intervals, vectors and matrices may be linked. For further details refer to [2].

3. **Verified Computation of the Procedure Error via Automatic Differentiation**

The only difficulty in computing the procedure error is the definition of a higher derivative of f for an interim point $\xi \in [a,b]$. Since ξ is unknown, an inclusion of $f^{(j)}(\xi)$ is computed by replacing ξ by $[a,b]$. A numerical approximation for the computation of the j-th derivative is not suitable for obtaining a verified result. As to symbolic differentiation, the necessary effort is too high. In 3.1, the mathematical basis of so-called "automatic differentiation algorithms" is specified as numerically validating the values of arbitrarily high derivatives at points or intervals. In 3.2 implementation is considered.

3.1 Automatic Differentiation Algorithms

Employing recursive computations, the method of "automatic differentiation" [2], [4], [9], [10], [11] provides inclusions of Taylor coefficients or derivatives. The algorithms are valid for real- or interval-valued points. Let u, v, w be real functions, which are analytical in an environment of the position t_0 (m times differentiable with a sufficiently high m also suffices).

The Taylor coefficients $(u)_k$ of a function u are

(1a) $(u)_k := \frac{1}{k!} \cdot u^{(k)}(t_0) := \frac{1}{k!} \cdot \frac{d^k u}{dt^k}(t_0),\quad \text{for}\quad k \geq 0$

or

(1b) $(u(\tau))_k := \frac{1}{k!} \cdot u^{(k)}(\tau),\quad \tau \in \mathbb{R}\ \text{ or }\ \tau \in \mathbb{IR}.$

In this notation the Taylor series of u at t_0 is written

(2) $u(t) = \sum_{k=0}^{\infty} (u)_k \cdot (t - t_0)^k\ .$

Thus, for functions being composed of arithmetic operations of other functions, the following computing rules may be indicated at once, which are, however, merely rules for computing power series:

(3a) $(u \pm v)_k = (u)_k \pm (v)_k$

(3b) $(u \cdot v)_k = \sum_{j=0}^{k} (u)_j (v)_{k-j}$

(3c) $(u / v)_k = \frac{1}{v} ((u)_k - \sum_{j=1}^{k} (v)_j \cdot (u/v)_{k-j})$

Proof of (3c):

With $w := u/v$ it holds: $u = v \cdot w$. Applying (3b) results in:

$$(u)_k = \sum_{j=0}^{k} (v)_j (w)_{k-j} = (v)_0 (w)_k + \sum_{j=1}^{k} (v)_j (w)_{k-j}$$

$$= v \cdot (w)_k + \sum_{j=1}^{k} (v)_j (w)_{k-j} \quad .$$

Solving this equation with respect to $(w)_k = (u/v)_k$ results in (3c). □

Due to the trivial relations

(4a) $(c)_0 = c$, $(c)_k = 0$, for $k \geq 1$, c is a constant

(4b) $(t)_0 = t_0$, $(t)_1 = 1$, $(t)_k = 0$, for $k \geq 2$, for the independent
      variable t

the Taylor coefficients for arbitrarily rational functions of any order k
may be computed by means of recursive computing of all partial expressions
first for the coefficients for k = 0 and then for k = 1, etc.

A comparison of the coefficients of

(5) $u'(t) = \sum_{k=1}^{\infty} k \cdot (u)_k \cdot (t-t_0)^{k-1} = \sum_{k=0}^{\infty} (k+1) \cdot (u)_{k+1} \cdot (t-t_0)^k$

and

(6) $(u')_k = \frac{1}{k} \cdot (u')^{(k)} (t_0) = \frac{1}{k!} \cdot u^{(k+1)} (t_0)$

$$\implies u'(t) = \sum_{k=0}^{\infty} (u')_k \cdot (t-t_0)^k$$

implies

(7) $(u')_k = (k+1) \cdot (u)_{k+1}$ or $(u)_{k+1} = \frac{1}{k+1} \cdot (u')_k$.

Due to (7), the chain rule may be applied, and thus, computation formulas for the Taylor coefficients analogous to those of (3) may be derived for a far larger class of functions:

(8) $w(t) = v(u(t)) \implies w'(t) = v'(u(t)) \cdot u'(t)$.

Similar to proof (3c), by means of (3b) and (7), a recursion formula for $(w)_k$ is derived being illustrated by means of the following exponential function:

$$w = e^u \implies w' = e^u \cdot u' = w \cdot u'$$

$$\implies (w')_{k-1} = (w \cdot u')_{k-1}, \quad k \geq 1.$$

From (3) and (7) it follows that

$$k \cdot (w)_k = (w')_{k-1} = (w \cdot u')_{k-1} = \sum_{j=0}^{k-1} (w)_j \cdot (u')_{k-1-j} =$$

$$= \sum_{j=0}^{k-1} (w)_j \cdot (k-j) \cdot (u)_{k-j},$$

and, therefore, from $(w)_k = (e^u)_k$,

(9) $(e^u)_k = \sum_{j=0}^{k-1} \left(1 - \frac{j}{k}\right) \cdot (e^u)_j \cdot (u)_{k-j}$, $\quad k \geq 1$.

In a similar manner, recursion formulas for all other ordinary analytical standard functions are derived. Some of them are illustrated as examples without proof (for $k \geq 1$):

$$\begin{cases}
(u^a)_k & = \frac{1}{ku} \cdot \sum_{j=0}^{k-1} (a(k-j)-j)(u)_{k-j}(u^a)_j , \quad a \in \mathbb{R} \text{ constant,} \\[3mm]
(\ln u)_k & = \frac{1}{u} \cdot ((u)_k - \sum_{j=1}^{k-1} (1-\frac{j}{k})(u)_j (\ln u)_{k-j}) \quad , \\[3mm]
(\sin u)_k & = \frac{1}{k} \cdot \sum_{j=0}^{k-1} (j+1)(\cos u)_{k-j-1}(u)_{j+1} \quad , \\[3mm]
(\cos u)_k & = -\frac{1}{k} \cdot \sum_{j=0}^{k-1} (j+1)(\sin u)_{k-j-1}(u)_{j+1} \quad , \\[3mm]
(\sinh u)_k & = \frac{1}{k} \cdot \sum_{j=0}^{k-1} (j+1)(\cosh u)_{k-j-1}(u)_{j+1} \quad , \\[3mm]
(\cosh u)_k & = \frac{1}{k} \cdot \sum_{j=0}^{k-1} (j+1)(\sinh u)_{k-j-1}(u)_{j+1} \quad , \\[3mm]
(\arctan u)_k & = \frac{1}{1+u^2} \cdot ((u)_k - \frac{1}{k} \cdot \sum_{j=0}^{k-1} j(\arctan u)_j (1+u^2)_{k-j}) .
\end{cases}$$

(10)

With sin and cos (sinh and cosh also), it is obvious that the recursion formulas must be applied in pairs, and with the arctan, the Taylor coefficients of $1 + u^2$ still must be computed.

In the differentiation package specified in section 3.2, the computation of the Taylor coefficients are realized for the dyadic operations described above (exponentiation included) and the following functions:

- ln, exp, u^2, $\sqrt{u}$,
- the 4 trigonometrical functions with their inverses,
- the 4 hyperbolic functions with their inverses.

Hence, the computation of the derivative $u^{(k)}(\tau)$ follows trivially from (1) with respect to

(11) $\quad u^{(k)}(\tau) = k! \cdot (u(\tau))_k \quad .$

3.2 Implementation of the Differentiation Package

Now the arithmetic rules and recursion formulas described in section 3.1 are to be implemented in the programming language PASCAL-SC. For a data type, we choose a dynamic array. The components represent the Taylor coefficients.

type itaylor = dynamic array [*] of interval;

This version serves for the verified computation of Taylor coefficients, such that we obtain intervals for results. The implementation of a package computing real approximations for the derivatives was omitted, since it is was not considered necessary for this paper. For a clearer understanding of important items some details are simplified in this description.

The structure of the differentiation package is illustrated in the following simple example.

Let it be given that $f(x) := x^2 - 3x + 1$

From rules (3) and (4) with ubound = 3, we obtain the following statement:

The independent variable x is defined by

var x:itaylor [0..3];

In this description according to (4), the variable x and the constants 3 and 1 are illustrated by:

$$x \longrightarrow (x, 1, 0, 0)$$
$$3 \longrightarrow (3, 0, 0, 0)$$
$$1 \longrightarrow (1, 0, 0, 0)$$

(3) applied to x^2 (= x · x) gives:

$$x^2 \longrightarrow (x^2, 2x, 1, 0)$$

From (3a), we obtain for $f(x_0)$:

$$f(x_0) \longrightarrow (x_0^2 - 3x_0 + 1, \; 2x_0 - 3, 1, 0) = ((f)_0, (f)_1, (f)_2, (f)_3)$$

With $f''(x_0) = 2! \cdot (f(x_0))_2$ we obtain:

$f'(x_0) = 2 \cdot 1 = 2.$

If x is initialized with $x_0 := 1$ or $[0,1]$, the result for the components of $f(x)$ is:

$f(x)$ $\lceil i \rceil$	$x_0 = 1$	$x_0 = \lceil 0,1 \rceil$
$i = 0$	-1	$[-2,2]$
$i = 1$	-1	$[-3,-1]$
$i = 2$	2	2
$i = 3$	0	0

In Table 1, all implemented operators are listed. The operand set Ω and the operand types K, R, I and "it" denote the following:

K : integer,   R : real,   I : interval,   it : itaylor,
$\Omega := \{ +, -, *, / \}$

```
± it, K ∘ it, it ∘ K, I ∘ it, it ∘ I, it ∘ it
for all ∘ ∈ Ω ,
K to_the K, K to_the R, R to_the K, R to_the R,
I to_the K, K to_the I, I to_the I,
I to_the R, R to_the I,
it to_the K, it to_the R.
```

Table 1: **itaylor**-operators

Table 2 lists all implemented functions with argument and result type itaylor. Since in PASCAL-SC an overloading of function names is possible, the function names are the same as for real or interval arguments.

```
sqr, sqrt, ln, exp,
sin, cos, tan, cot,
arcsin, arccos, arctan, arccot,
sinh, cosh, tanh, coth,
arsinh, arcosh, artanh, arcoth.
```

Table 2: **itaylor**-functions

Finally, let us give some examples for the implementation of an operator or a function as well as for a simple application in a main program. For further details refer to [2].

Example 1: Multiplication of two itaylor variables

```
global operator * (A,B:itaylor) res: itaylor [0..ubound(A)];
var    k,j: integer;
begin  for k:= 0 to ubound(A) do
         res[k] := ## (for j:= 0 to k sum (A[j]*B[k-j]));
end;
```

Example 2: The exponential function $y(u(x)) := e^{u(x)}$

```
global function exp(x:itaylor) : itaylor [0..ubound(x)];
var   k,j : integer;
      h   : itaylor [0..ubound(x)];
begin h[0] := exp(x(0));
      for k := 1 to ubound(x) do
    begin h[k] := null;
          for j:=0 to k-1 do h[k]:=h[k]+(k-j)*h[j]*x[k-j];
            h[k] := h[k]/k;
    end;
    exp := h;
end;
```

Example 3: Computing the third derivative of the function

$$f(x) = x * e^{1-x^2} \quad \text{at input points.}$$

```
program  example 3(input, output);
use   irari, itaylor;
var   i,j : integer;
      h,x : itaylor[0..3];
  x0 :   interval;
function f(x:itaylor):itaylor[0..ubound(x)];
begin   f:= x * exp(1-sqr(x));
end;
begin    (*Main program*)
      read(input, x0);
      expand(x,x0); h := f(x);
      write (output,  h[3] * 6);
end.
```

4. Numerical Quadrature via Modified Romberg-Extrapolation

4.1 The Romberg-Integration

We briefly introduce the conventional Romberg-Integration [12]. In consequence of previous calculations in chapter 2, the variables Q and R are requested in the formula:

$$J = Q + R = \int_a^b f(x)\,dx$$

In the T-table we choose for Q the element T_{mm} (in the following Q is often denoted analogously with T_{mm} or Q_{mm}):

$$(1) \quad \begin{matrix} T_{00} & & & \\ T_{10} & T_{11} & & \\ T_{20} & T_{21} & T_{22} & \\ \cdot & \cdot & \cdot & \\ \cdot & \cdot & \cdot & \\ \cdot & \cdot & \cdot & \\ T_{m0} & T_{m1} & \ldots & T_{mm} \end{matrix}$$

The elements T_{i0} are defined as follows:

$$(2) \quad T_{i0} := h_i \cdot \sum_{j=0}^{2^i} {}'' f(x_{ij})$$

with

$$(3) \quad h_i := \frac{b-a}{2^i}, \quad x_{ij} := a + j h_i, \quad j = 0(1)2^i, \quad i = 0(1)m$$

The double primes after the summation sign Σ in (2) signify that the first and the last term of the sum have to be bisected respectively. Thus, the elements of the first column, corresponding to the trapezium rule, are to be computed first. The element T_{ik} is computed by means of the two elements $T_{i,k-1}$ and $T_{i-1,k-1}$ lying to the left of T_{ik}, according to

$$(4) \quad T_{ik} := \frac{4^k \cdot T_{i,k-1} - T_{i-1,k-1}}{4^k - 1}, \quad 1 \le k \le i \le m.$$

Thus, the diagonal element T_{mm} follows recursively from (2) and (4). However, all elements of the T-table above and to the left of T_{mm} have to be computed!

From [12], concerning the error R, the following error formula holds:

$$(5) \quad R = - (b-a) \cdot \prod_{i=0}^{m} h_i^2 \cdot \frac{B_{m+1}}{(2m+2)!} \cdot f^{(2m+2)}(\xi), \quad \xi \in [a,b]$$

B_{m+1} is the Bernoulli-number which is computed via the following recursion:

$$(6) \quad B_k := (-1)^{k+1} \cdot B_{2k}(0),$$

with $B_{2k}(x)$ being the Bernoulli-Polynomial according to

$$(7) \quad \begin{cases} B_0(x) = 1 \\ B_k'(x) = k \cdot B_{k-1}(x) \quad , \quad k \geq 1 \\ \int_0^1 B_k(x)dx = 0 \quad , \quad k \geq 1 \end{cases}$$

The first Bernoulli-numbers read as follows:

$$B_1 = \frac{1}{6} , \quad B_2 = \frac{1}{30}, \quad B_3 = \frac{1}{42}, \quad B_4 = \frac{1}{30}$$

From (3),

$$h_0 = b-a , \quad h_i = h_0 \cdot 2^{-i}$$

thus

$$\prod_{i=0}^{m} h_i^2 = h_0^{2m+2} \cdot 2^{-m(m+1)}$$

Hence, R implies

$$(5') \quad R = \frac{-h_0^{2m+3} \cdot B_{m+1}}{2^{m(m+1)} \cdot (2m+2)!} \cdot f^{(2m+2)}(\xi) , \quad \xi \in [a,b].$$

There are difficulties in computing R only with respect to computing the derivative. The number ξ is unknown and replaced by $[a,b]$. Thus we obtain an inclusion $\hat{R}$ of R. Now the derivative $f^{(2m+2)}([a,b])$ is automatically computed as described in chapter 3. According to (6) and (7) the Bernoulli-numbers are recursively computed or stored in a table up to the maximum value that is necessary (cf. 4.2).

To achieve an arbitrarily requested accuracy, a good estimator for the total error is necessary. It is advisable first to compute Q approximatively and then to break off when the relative error between two successive diagonal elements of the T-table becomes smaller than a given errorbound. At this point, Q and $\hat{R}$ are validated numerically. If the interval diameter is still too large it may be reduced by bisecting the interval [a,b] or by advancing in the T-table. Since computing higher derivatives from a certain step in the T-table onwards takes more time and efforts than "bisecting" the interval, we do not exceed a maximum step in the T-table. As far as the requested adaptive procedure is concerned (see chapter 5), it would be better, however, to compute a verified enclosure of J immediately, whereas proceeding in the T-table diagonal as well as bisection etc. are subject to the validation of J and to drop the evaluation of approximations.

The modified Romberg-procedure, which is specified below, replaces the recursive computation of all T_{ik} by a direct evaluation of one scalar product.

4.2 The Modified Algorithm

The principal statement of the modification is defined as Theorem 1. Each T_{ik} may be computed as a linear combination of the functional values at the points x_{ij} with the weights w_{ikj}, after addition resulting in h_0 and after being computed recursively according to a similar formula as in the case of the T_{ik} in (4). Since all weights but the constant factor h_0 are independent of f, a, b, they may be stored in a table so that a fast direct computation of T_{mm} is allowed without computing one single T_{ik} beforehand! The proof of the first part of the Theorem implies simultaneously the construction rule for a recursive computation of the weights.

Theorem 1

---

1. Each T_{ik} is represented according to (4) as

   (8a) $\quad T_{ik} = \sum\limits_{j=0}^{2^i} w_{ikj} \cdot f(x_{ij}), \quad x_{ij} := a + j \cdot h_i,$

   $\qquad j = 0(1)2^i \quad , \quad 0 \le k \le i \le m$

   with

   (8b) $\quad \sum\limits_{j=0}^{2^i} w_{ikj} = h_0$

2. The weights w_{ikj} are computed for an arbitrary i as follows:

   (9) $\quad w_{i_0 j} = \begin{cases} h_0/2^{i+1} & , \quad \text{for } j = 0 \text{ and } 2^i \\ h_0/2^i & , \quad \text{else} \end{cases}$

   and for a fixed i with $1 \le k+1 \le i$ due to the recursion rule

   (10) $\quad \begin{cases} w_{i,k+1,2j} = \dfrac{4^{k+1} \cdot w_{ik,2j} - w_{i-1,k,i}}{4^{k+1} - 1} & , \quad j = 0(1)2^{i-1} \\[4mm] w_{i,k+1,2j+1} = \dfrac{4^{k+1} \cdot w_{ik,2j}}{4^{k+1} - 1} & , \quad j = 0(1)2^{i-1}-1 \end{cases}$

   for $\quad 1 \le k+1 \le i \le m.$

Proof:

    a)   for T_{ik} with $k = 0$ and i arbitrarily : direct

    b)   for T_{ik} with i fixed, $1 \le k \le i$: by complete induction

ad a)

It holds

$$T_{i0} = h_i \cdot \sum\limits_{j=0}^{2^i} {}''f(x_{ij}) = \frac{h_0}{2^i} \cdot \sum\limits_{j=0}^{2^i} {}''f(x_{ij}) =$$

$$= \frac{h_0}{2^i} \cdot \{ \sum\limits_{j=1}^{2^i-1} f(x_{ij}) + \tfrac{1}{2}f(a) + \tfrac{1}{2}f(b)\}$$

hence, it follows that

$$\sum_{j=0}^{2^i} w_{i0j} = \frac{h_0}{2^i} \cdot \left\{ 2^i - 1 + \frac{1}{2} + \frac{1}{2} \right\} = h_0 \qquad\qquad \text{q.e.d.}$$

and

$$w_{i0j} = \begin{cases} h_0/2^{i+1} & , \quad \text{for } j = 0 \text{ and } 2^i \\ h_0/2^i & , \quad \text{else} \end{cases} .$$

Therefore, (9) is proved by the second part of the Theorem.

ad b)

For T_{i0} the statement according to "a)" is valid.

By induction the statement is valid for all $T_{\tilde{i}\tilde{k}}$ with $0 \le \tilde{i} \le i$ and $0 \le \tilde{k} \le k$.

$T_{ik} \rightarrow T_{i,k+1}$:

$$T_{i,k+1} = (4^{k+1} - 1)^{-1} \cdot (4^{k+1} \cdot T_{ik} - T_{i-1,k}) =$$

$$= (4^{k+1} - 1)^{-1} \cdot \left(4^{k+1} \cdot \sum_{j=0}^{2^i} w_{ikj}\, f(x_{ij}) - \sum_{j=0}^{2^{i-1}} w_{i-1,k,j}\, f(x_{i-1,j}) \right)$$

With

$$x_{i-1,j} = a + j \cdot \frac{h_0}{2^{i-1}} = a + (2j) \cdot h_i = x_{i,2j}$$

it holds that

$$T_{i,k+1} = (4^{k+1} - 1)^{-1} \cdot \left\{ 4^{k+1} \cdot \sum_{j=0}^{2^{i-1}} w_{i,k,2j} \cdot f(x_{i,2j}) \right.$$

$$+ 4^{k+1} \cdot \sum_{j=1}^{2^{i-1}} w_{ik,2j-1} \cdot f(x_{i,2j-1})$$

$$\left. - \sum_{j=0}^{2^{i-1}} w_{i-1,k,j} \cdot f(x_{i,2j}) \right\}$$

$$(*)$$

$$= \sum_{j=0}^{2^i-1} (4^{k+1}-1)^{-1} \cdot (4^{k+1} \cdot w_{i,k,2j} - w_{i-1,k,j}) \cdot f(x_{i,2j})$$

$$+ \sum_{j=1}^{2^i-1} (4^{k+1}-1)^{-1} \cdot 4^{k+1} \cdot w_{i,k,2j-1} \cdot f(x_{i,2j-1})$$

$$(**)$$

where $(*)$ is $w_{i,k+1,2j}$ and $(**)$ is $w_{i,k+1,2j-1}$. Thus, for the sum of the weights it follows that

$$\sum_{j=0}^{2^i} w_{i,k+1,j} = \sum_{j=0}^{2^i-1} w_{i,k+1,2j} + \sum_{j=1}^{2^i-1} w_{i,k+1,2j-1} =$$

$$= \sum_{j=0}^{2^i-1} (4^{k+1}-1)^{-1} \cdot (4^{k+1} \cdot w_{i,k,2j} -$$

$$- w_{i-1,k,j}) + \sum_{j=1}^{2^i-1} (4^{k+1}-1)^{-1} \cdot 4^{k+1} \cdot w_{i,k,2j-1}$$

$$= (4^{k+1}-1)^{-1} \cdot 4^{k+1} \cdot \left(\sum_{j=0}^{2^i-1} w_{i,k,2j} + \sum_{j=1}^{2^i-1} w_{i,k,2j-1} \right) -$$

$$\sum_{j=0}^{2^i-1} (4^{k+1}-1)^{-1} \cdot w_{i-1,k,j}$$

$$= (4^{k+1}-1)^{-1} \cdot 4^{k+1} \cdot \sum_{j=0}^{2^i} w_{ikj} - (4^{k+1}-1)^{-1} \cdot h_0$$
$\uparrow$

due to induction assumption

$\downarrow$

$$= (4^{k+1}-1)^{-1} \cdot 4^{k+1} \cdot h_0 - (4^{k+1}-1)^{-1} \cdot h_0 = h_0 \quad .$$

$(*)$ and $(**)$ provide the recursion rule for computation of the weights w_{ikj}; thus (10) of the second part of Theorem 1 is proven. $\square$

Contrary to a recursive computation of T_{ik} where each time many function values have to be computed, only rational numbers according to (9) and (10) have to be defined. Since the weights are independent of f, fixed tables may be indicated so that a recursive computation is only necessary if the requested T-Table element exceeds the fixed table. This will not be

necessary in practice, since we do not exceed a maximum step in the T-table
(as already mentioned in 4.1). So, a scalar product is constructed out of
the weight vector

(11a) $\vec{w}_{ik} := (w_{ik0}, \ldots , w_{ikn})$

and the function vector

(11b) $\vec{f}_{ik} := (f(x_{i0}), \ldots , f(x_{in}))$

with

$$n := 2^i$$

such that Q is computed according to

(11c) $Q_{ik} = \vec{w}_{ik} * \vec{f}_{ik}$.

This procedure is faster and more accurate than the conventional one.

4.3 Computing the Weights

A computation of the weights w_{ikj} was carried out definitely up to the
element T_{55}. These were stored as rational numbers, i.e. as pairs of
integers. In consequence, a higher accuracy was achieved in computing Q.

All w_{ikj} for a vector $\vec{w}_{ik}$ were reduced to a common main denominator which
was balanced with h_0, etc. Some values are given as an example:

$$\vec{w}_{00} = \frac{h_0}{2}(1,1), \quad \vec{w}_{10} = \frac{h_0}{4}(1,2,1),$$
$$\vec{w}_{33} = \frac{h_0}{5670}(217, 1024, 352, 1024, 436, 1024, 352, 1024, 217).$$

4.4 Computing Effort and Rounding Errors

In comparison with the modified procedure (β), the conventional procedure
(α) requires greater computing effort and implies higher rounding errors.
To compute T_{mm} by (α),

$$n_\alpha = \frac{(m+2)(m+1)}{2}$$

T-table elements must be calculated, whereas to do the same by (β) only

$$n_\beta = 1$$

elements are necessary! The computation of the diagonal elements T_{00} to $T_{m-1,m-1}$, is, however, necessary, if the information is lacking. For example, the starting element should be a T_{ii} with $i > 0$ to achieve the requested accuracy (see chapter 5.2, Procedure "T4T5"!). If the most unfavourable case is chosen, then

$$n_\beta' = (m+1)$$

T-table elements must be computed. To obtain a verified result everything must be computed in intervals via (α). By (β), the diagonal elements may be computed directly which results in a great amount of time saved and increased accuracy. While T_{m0} is being computed via (α), T_{mm} has already been verified numerically via (β). In case of (α) compared to (β), rounding errors may accumulate faster due to the recursive computation or may result in a more extensive blow up for interval-computing.

4.5 The Algorithm 1

To reduce rounding errors, there exist different possibilities to combine bisection methods and methods of progression in the T-table. Advancing to the right in the T-table increases the effort by a largescale computation of the $f^{(2m+2)}([a,b])$, since all lower derivatives are required (see chapter 3). Advancing downwards increases the effort by enlarged function evaluations owing to the rise in the number of points. Unfortunately, you never know in advance which possibility is more favourable as to runtime and accuracy. The adaptive procedure described in chapter 5 shows how the additional time and effort mentioned above may be avoided in both cases. The following Algorithm 1 represents the coarse procedure which is the basis for the refined adaptive procedure of Algorithm 2 in chapter 5.

Algorithm 1:   Structure of the simple procedure

Step 1:   input of f, a, b, ϵ
$h_0 := b - a;$ $i := -1;$

Step 2: $i := i+1;$

a) Compute $\Diamond T_{ii}$ via (8), (9), (10),   i.e.:

$$T_{ik} = \sum_{j=0}^{2^i} w_{ikj} * f(x_{ij})$$

with the weights of the fixed, stored table
which are computed according to (9), i.e. due to

$$w_{i0j} = \begin{cases} h_0/2^{i+1} & , \text{ for } j = 0 \text{ and } 2^i \\ h_0/2^i & , \text{ else} \end{cases}$$

and recursively due to (10), i.e. according to

$$\begin{cases} w_{i,k+1,2j} = \dfrac{4^{k+1} \cdot w_{i,k,2j} - w_{i-1,k,i}}{4^{k+1} - 1} & , \quad j = 0(1)2^{i-1} \\[3mm] w_{i,k+1,2j+1} = \dfrac{4^{k+1} \cdot w_{ik,2i}}{4^{k+1} - 1} & , \quad j = 0(1)2^{i-1-1} \end{cases}$$

for $1 \leq k+1 \leq i \leq m.$

b) Compute $\Diamond \hat{R}_{ii}$ according to (5'), i.e.:

$$\hat{R}_{ii} = \frac{-h_0^{2i+3} \cdot B_{i+1}}{2^{i(i+1)} \cdot (2i+2)!} f^{(2i+2)} ([a,b])$$

Step 3:   <u>if</u> (d $(\Diamond T_{ii} + \Diamond \hat{R}_{ii}) < \epsilon$) or $(i = i_{max})$

<u>then</u> go to step 4   <u>else</u>   go to Step 2

Step 4:   output of the result $J_i := \Diamond T_{ii} + \Diamond \hat{R}_{ii}$ and
the values $i, d(J_i).$

5. Faster Reduction of the Total Error via Adaptive Refinement

In chapter 5.1 we deal with the basic idea of adaptive refinement and consider the coarse structure of the adaptive algorithm. The various algorithm segments are described as per the top-down method in chapter 5.2. Chapter 5.3 deals with the data types and implementation details which are used.

5.1 Basic Idea

Since the procedure error reduces towards 0 only with $O(h^p)$ if $h < 1$, the integration domain $I^{(0)} := [a,b]$ is divided first into subintervals, I_i, with $d(I_i) \leq 1/2$, in such a manner that subsequent points might be represented exactly. Proceeding from the first subdivision, the modified Romberg procedure is applied to the subintervals. If the errors resulting from the T-table-element T_{55} in an interval I_i are still too large, the subinterval is bisected into two new subintervals by Algorithm 1 where the starting element is T_{44}. If the error in one of the two new subintervals is still too large, a new bisection takes place; Algorithm 1 is applied starting with T_{44}, etc. Hence, bisection is only used in cases in which it is required. Thus, in subintervals showing good results no precious runtime is wasted (see figure 1). Note that figure 1 is not true to scale.

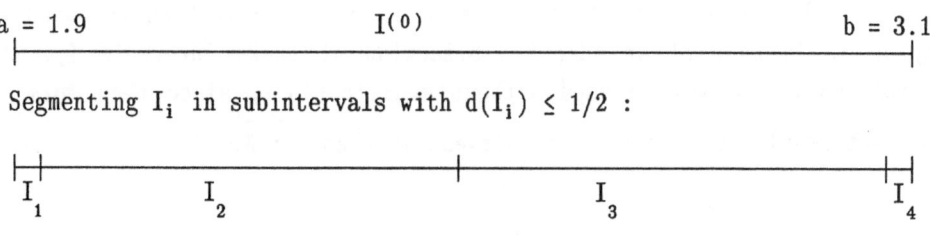

Segmenting I_i in further subintervals until the error is sufficiently small in these subintervals results in the following distribution:

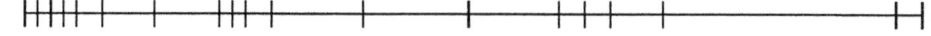

Figure 1:   Example for a "last" segmenting of the integration domain

<u>Remarks:</u>

I_1 is bisected once, $h_{min} = 0.05$

I_2 is bisected 5 times with a total of 10 subintervals, $h_{min} = 1/64$

I_3 is bisected 4 times with a total of 5 subintervals, $h_{min} = 1/32$

I_4 is not segmented, $h_{min} = 0.1$

Note that for the computation of $\hat{R}$ it holds that

$$(1) \quad \bigcup_{i=1}^{n} f^{(j)}(I_i) \subseteq f^{(j)}\left(\bigcup_{i=1}^{n} I_i\right).$$

Such inflation may be reduced by the computation of derivatives n-times. In general, this is not necessary and thus too expensive, since for the value of $\hat{R}$, it is not the relative diameter of $\hat{R}$, but the relative size of $\hat{R}$ compared to $\Diamond T_{mm}$ that is relevant for a good result. For example, let be $iabs(\hat{R}) < 10^{-14}$, if $iabs(\Diamond T_{mm}) \approx 1$ and if the computer provides 13 decimals. Then, it is absolutely sufficent to compute $\hat{R}$ exactly up to 1 digit so that for $J := \Diamond T_{mm} + \hat{R}$ the increase of $\hat{R}$ has only minimum effects. Therefore, $f^{(j)}$ is computed for the greatest possible j (here: $j = 2m+2$ with $m \leq 5$, such that $j = 12$) at the beginning of each interval I_i, $i = 1(1)k$. The lower derivative for each I_i has already been computed in chapter 3. In case of an adaptive bisection of these intervals I_i, the derivatives on I_i may be used without modifications, since they form an upper set resulting only in an irrelevant increase of $\hat{R}$.

5.2 Refinement Strategy and Adaptive Algorithm

First we illustrate a structure of the adaptive algorithm. Note that Step 3
is the most important and most complicated one.

Algorithm 2: Structure of the adaptive algorithm

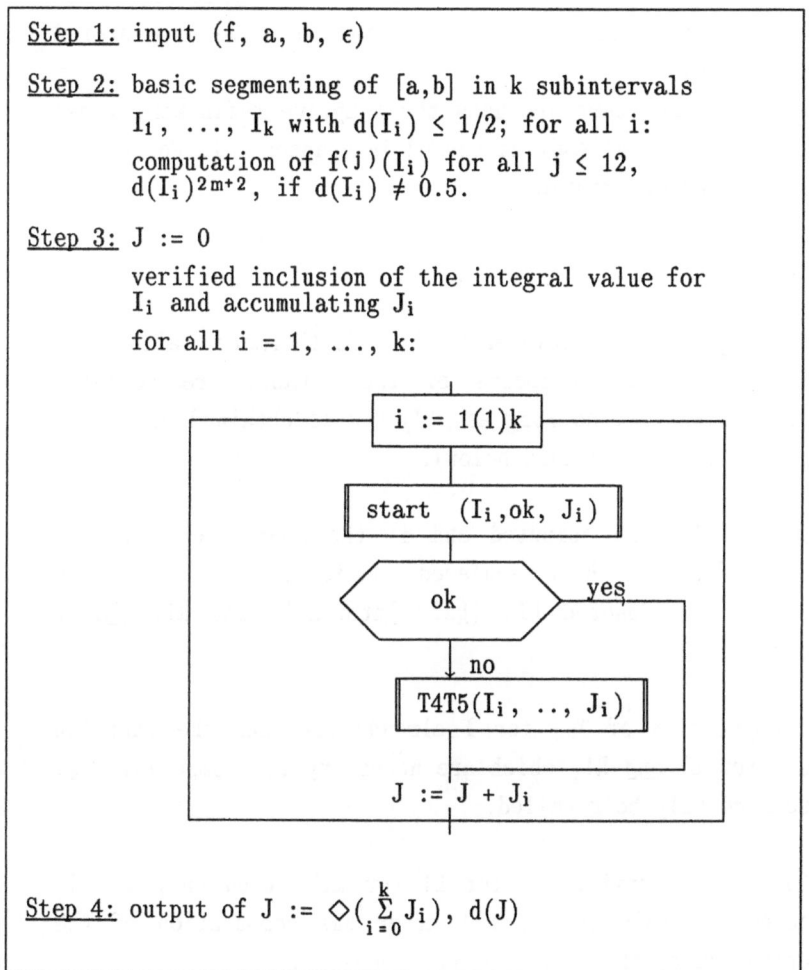

Step 1: input (f, a, b, ϵ)

Step 2: basic segmenting of [a,b] in k subintervals
I_1, ..., I_k with $d(I_i) \leq 1/2$; for all i:
computation of $f^{(j)}(I_i)$ for all $j \leq 12$,
$d(I_i)^{2m+2}$, if $d(I_i) \neq 0.5$.

Step 3: J := 0
verified inclusion of the integral value for
I_i and accumulating J_i
for all i = 1, ..., k:

i := 1(1)k

start (I_i, ok, J_i)

ok   yes

no

T4T5(I_i, .., J_i)

J := J + J_i

Step 4: output of J := $\diamondsuit(\sum_{i=0}^{k} J_i)$, d(J)

Remarks:
- ad step 1: ϵ is the requested relative diameter of the
  inclusion on J.
- ad step 2: the first interval I_1 := [a, x_1] and the last
  interval I_k := [x_k, b] are chosen with respect to x_1 and x_k

as an integer multiple of 0.5. Hence, the diameter of the intervals $I_i := [x_{i-1}, x_i]$ with $x_i := x_1 + i \cdot 0.5, i = 2(1)k-1$, is exactly $1/2$, which is advantageous for the subsequent computation of $\hat{R}$ as well as for bisections. Now, the derivations for the computation of $\hat{R}_{mm}$ are calculated for the supreme root of each tree and are used for further subnodes. Therefore, it is useful to perform this computation when dividing the basic interval and thus to have the stored values ready.

- ad step 3: This part includes the adaptive refinement whose fundamentals are already discussed in chapter 5.1. Following, is a detailed description.

Refinement strategy:

(1) Proceeding from a subinterval I_i (furthermore called I), sucessive diagonal elements of the T-table are computed starting with T_{00}, so long as $\rho(J) < \epsilon$ is obtained or T_{55} is computed ($\rho(J)$ is defined below).

(2) If in (1) T_{55} is computed and if the requested accuracy, however, has not been achieved, I is bisected. The new intervals are denoted LI, (Left Interval), and RI, (Right Interval).

(3) The computation of T_{55} for I already provides the function values for LI and RI, which are necessary for computing T_{44}. These need only be recalled.

(4) If T_{44} is not sufficient for LI (or RI), then only T_{55} is computed for this LI (or RI) after the computation of the function values which thus become necessary.

(5) If T_{55} for LI (or RI) is not sufficient in (4) we proceed to (2) for this LI (or RI) as long as the requested accuracy is sufficient or a pregiven maximum bisection number is achieved. Thus, only these LI (or RI) are further bisected which fulfill the condition that $\rho(J) > \epsilon$!

In the following, $\rho(J_{mm})$ denotes a measure for the relative or absolute error of J_{mm}, such that:

If $0 \in J_{mm}$,

then: $\rho(J_{mm}) := d(J_{mm})$

else: $\rho(J_{mm}) := d(J_{mm})/> \min \{|J_{mm}.\inf|, |J_{mm}.\sup|\}$

All T_{ii} are directly computed intervalwise by means of <u>one</u> optimal scalar product (see chapter 4).

Part (1) of the above refinement strategy is divided into a "start computation up to T_{33}", which is called up only <u>once</u> and computations for T_{44} and T_{55} which may occur more often. The corresponding procedure errors $\hat{R}_{ii}$ are computed as well. In the two following sections the two most important procedures are described. In "start" and "T4T5" further routines are called which are now presented briefly. T_{mm}, $\hat{R}_{mm}$, J_{mm} are computed as follows:

$$J_{mm} := T_{mm} + \hat{R}_{mm}$$

with

$$T_{mm} := \vec{W}_{mm} * \vec{f}_{mm} \qquad \text{according to (11c) from chapter 4.2}$$

and

$$\hat{R}_{mm} := \frac{-h_0^{2m+3} \cdot B_{m+1}}{2m(m+1) \cdot (2m+2)!} \cdot f^{(2m+2)}(I_i)$$

according to (5') from chapter 4.1.

Note how $\vec{f}_{mm}$ is computed and stored (see below), and how h_0^{2m+3} is computed after bisection, (h_0^{2m+3} is for almost all cases a power of 2). Even when not a power of 2, the "new h_0" (= h_{new}) may be computed by means of the "old h_0" (= h_{old}) after bisection according to $h_{new}^{2m+3} := (h_{old}^{2m+3})/2^{2m+3}$, so that it is only the two previously stored numbers which must be multiplied.

5.2.1 The Procedure "Start"

The procedure "start" computes the 1-th subinterval for the variables T_{oo} ... up to a maximum of T_{33}. As soon as the requested accuracy is obtained "ok" is set to true and $T_{jj} + \hat{R}_{jj}$ is stored in a chained list in which the index "start" points (see figure 5, chapter 5.3) are to be stored afterwards in the long accumulator. If ϵ is not achieved up to T_{33} the function values still necessary for T_{44} will be computed and transferred by F.

start $(1, \epsilon, \text{start}, F, \text{ok}):$

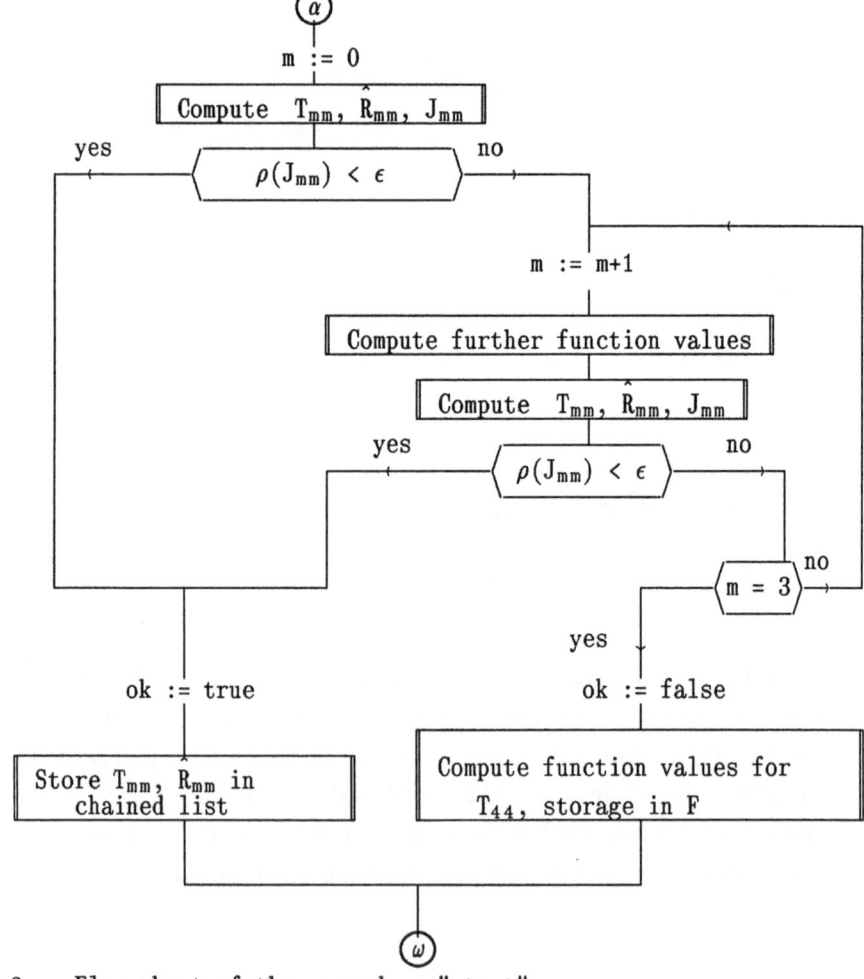

<u>Figure 2:</u> Flow chart of the procedure "start"

5.2.2 The Procedure "T4T5"

Since the procedure "T4T5" is only called if "ok" from "start" is false, it already contains in IV the function values which are necessary to compute T_{44}. For further function values, T_{55} is computed as needed. Everything refers to the interval $I := [LB, UB]$. If the requested accuracy is achieved at T_{44} or T_{55}, the result is stored as in "start" (see chapter 5.2.1). "last" shows the list element that was created last in the chained list of the T_{mm} and $\hat{R}_{mm}$ values. If this element is not achieved bisection takes place and - via transfer of the corresponding function values - this procedure is called recursively both for the interval LI and for RI. "reference" points to the actual node (see chapter 5.3).

T4T5 (reference, last, IV, LB, UB, level, ϵ, 1):

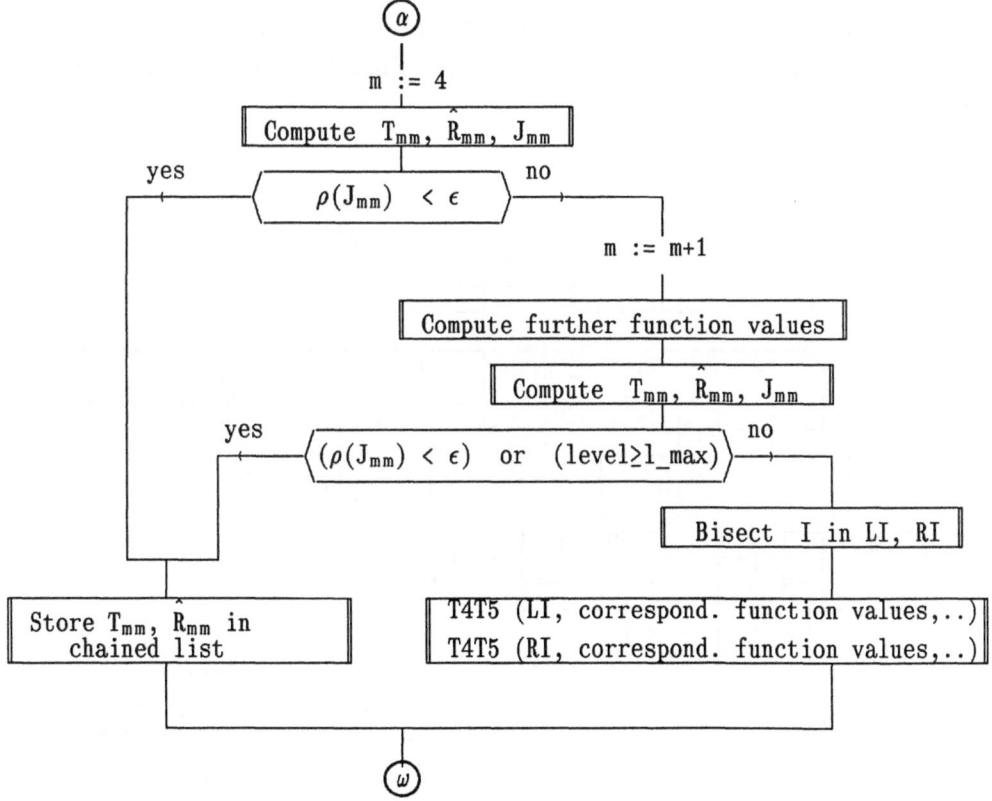

Figure 3:   Flow chart of the procedure "T4T5"

Remarks:

1. The procedure "T4T5" is programmed for LI and RI by a recursive call. To avoid an "endless loop", the procedure is regularly ended after a certain number of bisections, and the last J_{mm}-value is stored.

2. For different reasons, it is important to store the function values. Even if the requested accuracy is achieved for a certain interval, the sum of all subintegrals may lead to overestimations. Possibly, a further refinement might become necessary where the previously calculated function values are needed.

5.3 Applied Data Types and Implementation Details

Adaptive refinement of the subintervals I_i according to procedure T4T5 corresponds to the structure of a tree at which each node points to a left and a right son for the duration of the bisection. When a node has no sons, the computation is satisfactory and the function values and the integral values have to be stored or administrated. Figure 4 shows an example for a simple tree.

Step

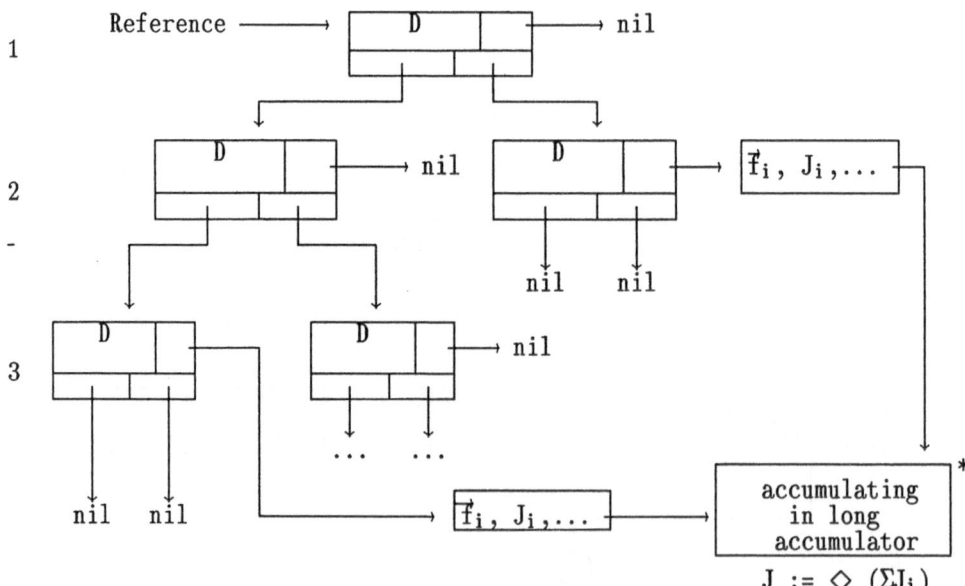

$$J := \diamond \ (\Sigma J_i)$$

Figure 4: Example for a Tree for Adaptive Segmenting

<u>Remarks:</u>

1. A node has the following elements:

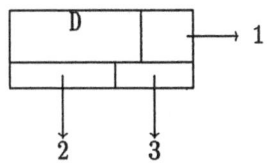

D: The data section contains the following
    (1) left and right bound of the interval belonging to this
        node.
    (2) Step of the tree in which the node lies.

1: A pointer which points to, provided the accuracy is sufficient,
    - an array with the function values

    - the integral value, distinguished between T_{mm} and $\hat{R}_{mm}$
    Otherwise, it points to nil.

2 and 3: pointers, which, provided accuracy is not sufficient, point

to

    nodes representing the left or right part of the bisected interval
    (left and right son).
    Otherwise they point to nil.

2. The function values are only stored at the leaves, if an interval has
    been "accepted". Storage space is only provided as it becomes
    necessary. If the step "l_max" is achieved, <u>no</u> further division should
    take place. Hence, maximum bisection is (l_max - 1) times.

3. Accumulation of J_i (see figure 4 at "*") is done as follows: A vector
    of the length k_max (= maximum number of trees) is created whose
    components are pointers of a chained list. The list elements are
    distinguished between T_{mm} and $\hat{R}_{mm}$ to minimize inflation when adding.
    All T_{mm} and $\hat{R}_{mm}$ of all list elements of each tree are accumulated
    intervalwise in the long accumulator so that only one rounding is
    required. Figure 5 illustrates the structure of the required
    "partialresultvector" (see below). k denotes the number of definitely
    existing trees.

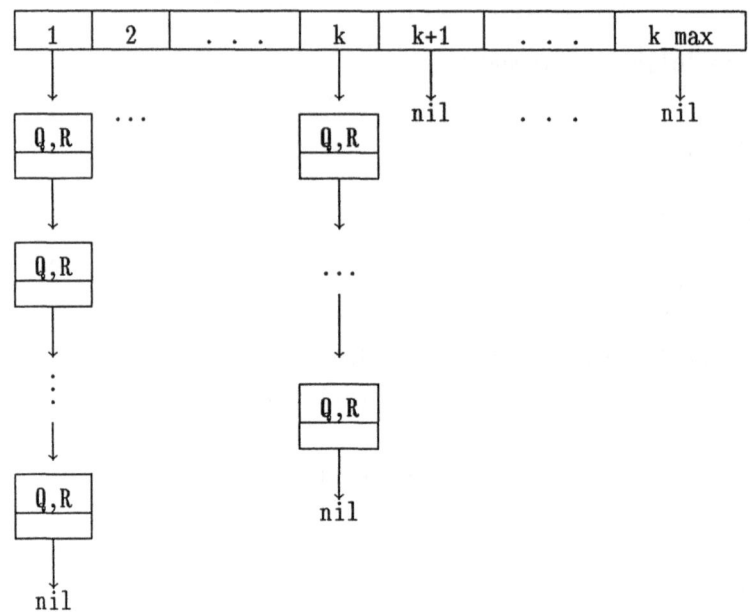

Figure 5: Structure of the data type "partialresultvector"

To implement the above data structure in PASCAL-SC the following data types are defined:

```
const       l_max = 6;(* max. levelnumber *)
            k_max = ...; (* e.g. 22 → d([a,b]) ≤ 10 *)
type        fvaluepointer = ↑fvaluenode;
            fvaluenode = record fvalue_array : array [0..32] of interval;
                                Q, R          : interval;
                     end;
            pointer   = ↑node;
            node      = record  LGR, RGR : real;
                                step     : integer;
                                fvalue   : fvaluepointer;
                                LS, RS   : pointer;
                     end;
            subinterval_array = array [1..k_max] of pointer;
            ivaluepointer = ↑ivaluelist;
            ivaluelist = record  Q,R : interval;
                                 subs: ivaluepointer;
                     end;
            partialresultvector = array [1..k_max] of ivaluepointer;
```

The implementation of the algorithms described above are now explained in detail. Step 2 of algorithm 2, segmenting the basic interval and computing the derivatives, is amplified in the flow chart of figure 7, where the definition of the first subintervals I_m is made possible by a separate procedure (see figure 6).

The upper boundary k_max for the number of subintervals is e. g. 22, provided that $d([a,b]) \leq 10$ is allowed. Thus a variable "sia" of the type "subinterval_array" contains a maximum of 22 trees documenting the adaptive refinement and possessing in the leaves all the information necessary for the solution of our problem.

Figure 8 shows the flow chart of the procedure "T4T5" and remarks on the storage of the necessary function values.

<u>Define I_m:</u>

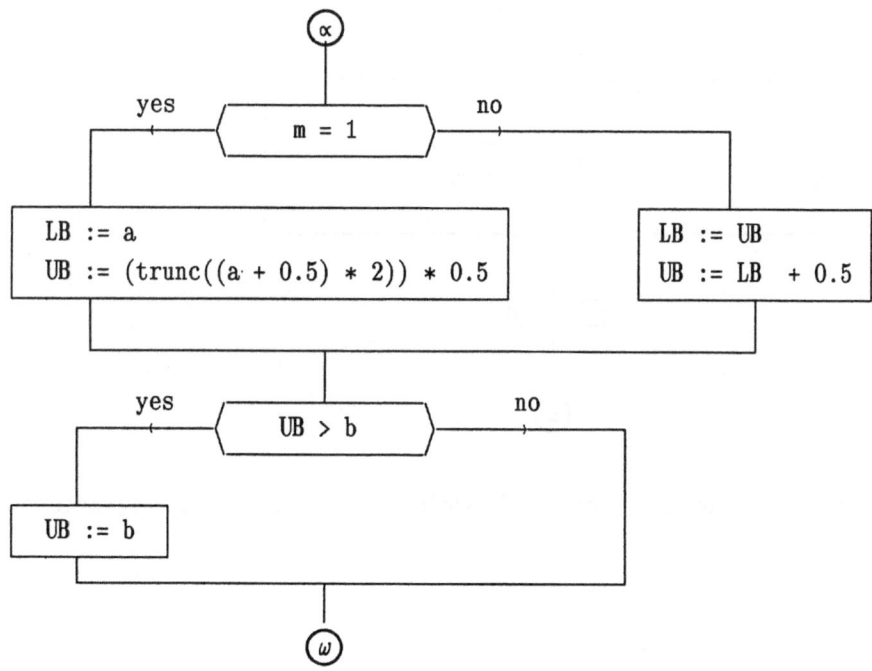

<u>Figure 6:</u>   Flow Chart of the Procedure "define_I_m"

procedure segmenting_and_derivation (var k:integer):

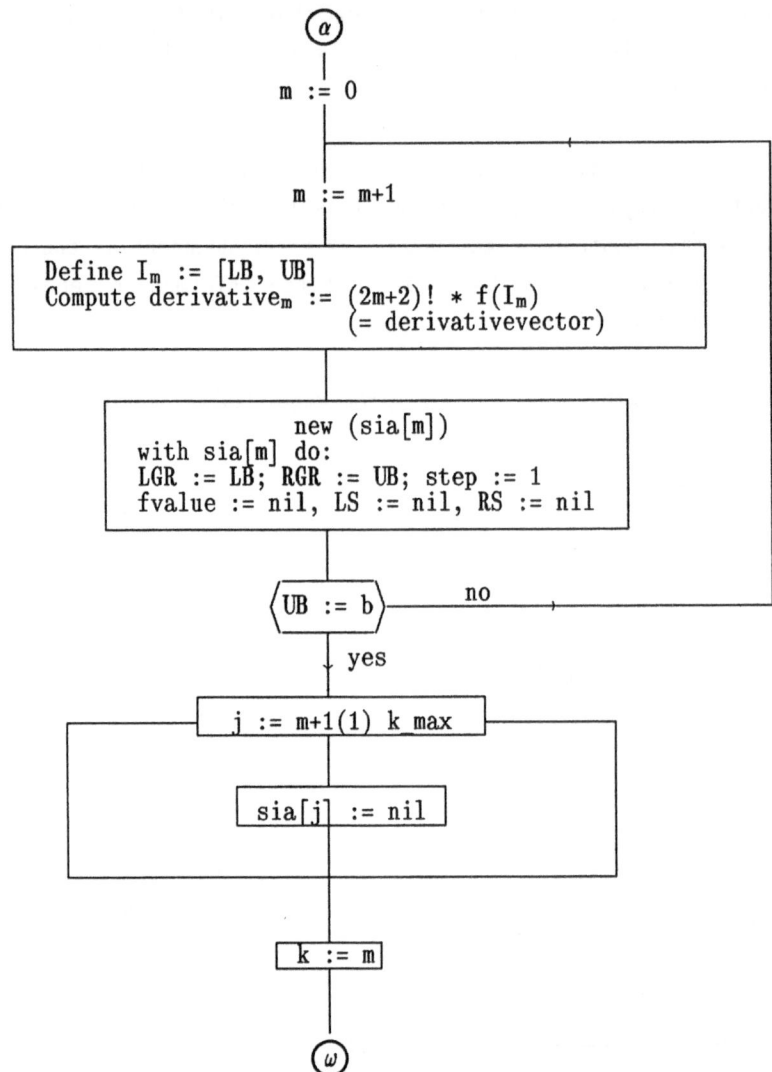

Figure 7:   Flow Chart of the Procedure "segmenting_and_derivation"

procedure T4T5 (reference: pointer IV: ivector; LB, UB:real; level:integer;
        ϵ : real);

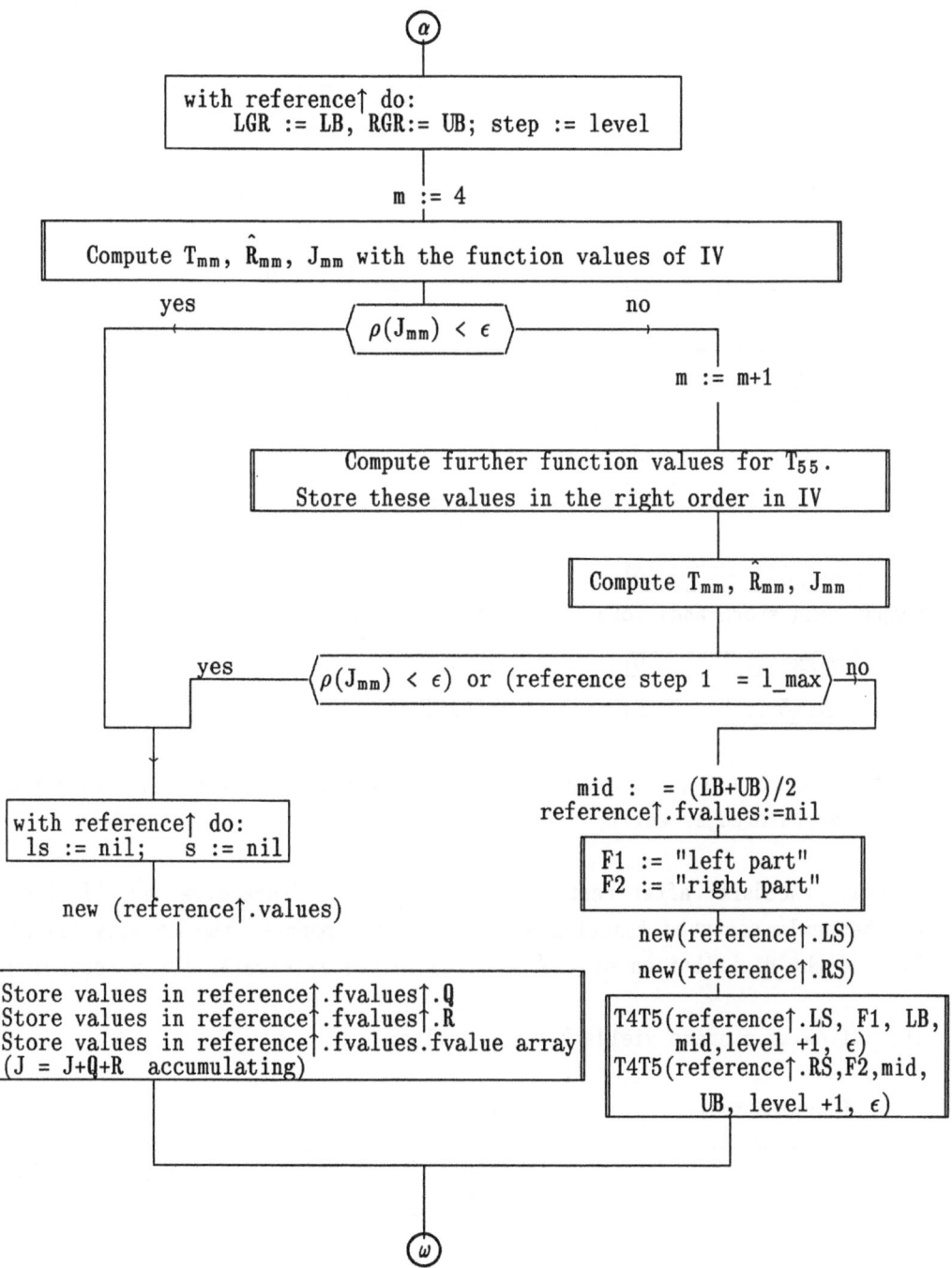

<u>Figure 8:</u>   Detailed Flow Chart of the Procedure "T4T5"

Remarks on the Storage of the Function Values in "start", "T4T5":

Computation of T_{00}:   x———————————————————x
                T_{11}:   x———————————x———————x
                T_{22}:   x————x————x————x————x
                etc.

        function values   weights
T_{00} 0 ...   32

    | xxo ..   0 | | xxo ..   0 |

restore:

    | xoxo ..   0 |

compute and store additional function values for T_{11}:

    | xxxo ..   0 |

T_{11}: | xxxo ..   0 | | xxxo ..   0 |

restore:

    | xoxoxo ..   0 |

compute and store additional function values for T_{22}:

    | xxxxxo ..   0 |

T_{22}: | xxxxxo ..   0 | | xxxxxo ..   0 | etc.

If an arbitrary T_{ii} is computed with sufficient accuracy, the function
values are stored as well.

General procedure: after restoring the zeros, the "vacancies" are "filled"
by the newly computed function values. In this manner, the computation of
T_{mm} and the distribution of the function values is clearer for a recursive
call.
The following storage yields better results:

        function values   weights

T_{00}: | xo .... ox | | xo .... ox |

T_{11}: | xo .. oxo .. ox | | xo .. oxo .. ox | etc.

This version is implemented, since on the one hand, the resulting dense
vectors are not at all adverse to the scalar product and, on the other
hand, not restoring zeros is an immense advantage.

6. Numerical Results

Finally, let us give some numerical results. Let be

$$\tilde{J} \approx J := \int_a^b f(x) \, dx \in \hat{J}, \quad \in \diamond J \, ,$$

where J denotes the exact value, $\tilde{J}$ an approximation and $\hat{J}$ or $\diamond J$ the inclusion which is computed by evaluation of the primitive function or by a verifying algorithm. Let eps be the requested relative error and rho the relative error resulting from the calculation of $\tilde{J}$ or $\diamond J$. So rho represents an estimation or inclusion of the error. "#f_i" denotes the number of function evaluations. A "*" in the column "Remarks" means that the quadrature delivers a better result than the primitive function does.

f(x)	[a,b]	eps	rho	#f_i	Remarks
sin(x)	$[0,\pi]$	1E-6	5E-9	33	
		1E-12	1.5E-12	71	
	[-0.1,0.1]	1E-9	2E-14	10	* , J = 0
$1/(25-x^2)$	[0,2]	1E-5	1E-6	12	
		1E-12	1.5E-12	44	
1/x	[1,2]	1E-11	2.4E-12	50	
	[0.1,1]	1E-11	1E-12	197	
	[0.01,1]	1E-11	2E-12	455	
e^{-x^2}	[0,5]	1E-11	2E-12	442	
	[0,1]	1E-11	9E-12	50	
$x-\ln(e^x)$	[0,1]	1E-12	8E-13	136	J = 0
	[11,12]	1E-11	2E-11	130	
$x-1+\exp(2\ln\sqrt{1+x})$	[1,2]	1E-11	3E-12	34	J = 3
$\sin^2 x + \cos^2 x$	[0,1]	1E-11	3E-12	34	J = 1
	[-5,5]	1E-11	3E-12	340	
cos x	[1.4,1.6]	1E-9	4E-12	10	*
	[0,6.28]	1E-11	8E-7	117	
e^x	[-1,1]	1E-9	2E-12	36	
$1/(1+x^2)$	[-1,1]	1E-12	3E-12	200	

Table 3: Numerical Results

Table 3 illustrates that for the most part the inclusion as well as the classical algorithms provide excellent inclusions or approximations. In a few examples, however, the classical algorithm (C) and the inclusion algorithm (I) do not agree, i.e. $\tilde{J} \notin \Diamond J$. Let be

$$J = \int_0^1 \frac{dx}{0.01 + (3x-1)^2} \quad ,$$

$\tilde{J}$, $\hat{J}$, $\Diamond J$, eps and rho as defined above and $\bar{J}$ an evaluation of the primitive function with 100-digit calculation. Thus we obtain

$$\hat{J} = 9.973\ 218\ 684\ 5^{91}_{83} \qquad \tilde{J} = 9.973\ 218\ 684\ 58896...$$

For eps = 1E-11 (I) and (C) provide:

$$\Diamond J = 9.973\ 218\ 684^{6031}_{5726} \implies d(\Diamond\hat{J}) \approx 4 \cdot d(\hat{J}) \qquad (\text{rho=3E-12, } \#f_i = 1296)$$

$$\tilde{J} = \underline{9.973\ 218\ 667}\ 569 \implies \text{dist } (\tilde{J},\ \Diamond J) \geq 2 \cdot 10^3 \cdot d(\Diamond J) = 1.7 \cdot 10^4 \cdot \text{rho}$$
$$(\text{rho=1E-12, } \#f_i = 1025)$$

The true relative error R is 1.7E-8. Thus, the error estimator leads us to believe that the error is about 17000 times smaller than it really is!

If we apply the same adaptive segmenting for (C) as we do for (I) and if the error estimator of (C) is still used as truncation criterion, then we receive for eps = 1E-11:

$$\tilde{J} = \underline{9.973\ 218\ 684\ 579} \implies \text{dist}(\tilde{J}, \Diamond J) = 4E-12 = 0.3 \cdot \text{rho}$$
$$(\text{rho=1.3E-11, } \#f_i = 402)$$

R = 1E-12 denotes a true error which is 10 times smaller than the error being calculated by the estimator. Hence, as far as this special example is concerned, we obtain a reliable result although it is highly overestimated. Table 4, however, shows that this is not always the case. In this table all three relations "rho $\cong$ R", "rho >> R" and "rho << R" emerge. Thus, the reliability of the error estimator in case of classical algorithms is rather doubtful. Even the improvement of (C) via adaptive refinement is not a sufficient means of obtaining reliable error estimations.

eps	1E-4	1E-6	1E-8	1E-10
rho	1E-5	4.5E-7	1.3E-9	9.57E-12
R	5E-4	2.2E-5	1.6E-9	6.5E-12

(R: true relative error)

Table 4: Doubtful Reliability of the Classical Error Estimator

As in case of many numerical problems, it is shown that an approximation procedure may provide true results only with high probability. If, however, guaranteed results as well as errorbounds for all (even the ill-conditioned) cases are requested, verified inclusion algorithms (see algorithms 2 in chapter 5) cannot be overlooked. Inclusion algorithms enable the user to calculate small bounds for the required integral in a fast and uncomplicated manner via automatic differentiation algorithms and transformation of the Romberg-Extrapolation into a direct dot product. Thus, a true error control has become possible via adaptive stepsize control!

7. <u>References</u>

[1] Alefeld, G., Herzberger, J.: Introduction to Interval Analysis. Academic Press, New York (1983)

[2] Bohlender, G., Rall, L.B., Ullrich, Ch., Wolff v. Gudenberg, J.: PASCAL-SC, A Computer Language for Scientific Computation, Academic Press, New York (1987)

[3] Corliss, G. F.: Computing Narrow Inclusions for Definite Integrals, published in [4]

[4] Kaucher, E., Kulisch, U., Ullrich, Ch. (Eds.): Computerarithmetic, Scientific Computation and Programming Languages. B. G. Teubner, Stuttgart (1987)

[5] Kaucher, E., Miranker, W.L.: Self-Validating Numerics for Function Space Problems. Academic Press, New York (1984)

[6] Kelch, R.: Kontinuierliche Einschließung der Lösungsfunktion funktionaler und parameterabhängige Probleme auf dem Computer, ZAMM 69, 4/5, to appear.

[7] Kulisch, U.W., Miranker, W.L. (Eds.): Computer Arithmetic in Theory an Practice. Academic Press, New York (1981)

[8] Kulisch, U.W., Miranker, W.L. (Eds.): A New Approach to Scientific Computation. Academic Press, New York (1983)

[9] Lohner, R.: Einschließung der Lösung gewöhnlicher Anfangs- und Randwertaufgaben und Anwendungen; Dissertation, Universität Karlsruhe (1988)

[10] Moore, Ramon E.: Interval Analysis. Prentice Hall, Englewood Cliffs, New Jersey (1966)

[11] Rall, L. B.: Optimal Implementation of Differentiation Arithmetic, published in [4]

[12] Stoer, J.: Einführung in die Numerische Mathematik I. Springer, Berlin (1979)

Solving Nonlinear Equations with Verification of Results

Diamond Deliverable D3-4

G. Schumacher
Universität Karlsruhe

Abstract:

This paper considers enclosure methods for the solution of systems of algebraic equations. Apart from this main interest of finding enclosures, questions of accuracy and the treatment of crucial cases are within the scope of the report. As part of the ESPRIT-project DIAMOND these questions were examined in the context of new concepts in computer arithmetic such as the availability of an exact scalar product and the embedding of this product in higher programming languages.

1. Introduction

In this paper, we consider a function $f: D \subseteq \mathbb{R}^n \to \mathbb{R}^n$ and we ask for the values $x \in D$ for which f disappears. In particular, we are interested in the isolated zeros of f, such that

$$(f(x) = 0) \wedge (\exists\ \epsilon > 0: f(y) \neq 0 \quad \forall y \in U_\epsilon (x) \setminus \{0\}) \qquad (1)$$

where $U_\epsilon(x)$ is an n-dimensional ball with center x and radius ϵ. Where necessary we shall make further restrictions on f such as continuousness, existence of partial differivatives, and so forth. Problems of form (1) are called systems of nonlinear equations since each component $f_i: D \to \mathbb{R}$ of f, $i = 1, \ldots, n$ represents an equation for the unknowns x_i, $i = 1, .., n$.

The numerical treatment of (1) has been discussed extensively in the past (see [Ort 70]). Since the exact solution is, in general, neither representable nor computable; it is of great interest to find bounds for the solutions. Evidence from studies [Ale 83], [Bau 87], [Kau 82] indicates that interval analysis is an appropriate way to compute bounds for the solution.

In the context of the DIAMOND project, this paper follows up five goals:

- It examines known methods for including solutions of (1) and develops some extensions.

- Because many enclosure methods are based on "classical" approximations, we study such methods for numerically sensitive areas and look for improvements.

- Apart from numerical difficulties there are mathematical difficulties in terms of the iterative nature of all appropriate procedures. Since, in general, only local convergence is available, we are forced to find good initial guesses that suffices one needs.

- A "DIAMOND Nonlinear System Solver" should provide correct solutions even in critical cases. Yet, this should not lead to naive use of such a tool. Since numerical problems are often connected with problems in the associated physical model, the user should also be able to get all background information he wants to know.

- Easy handling such as a comfortable I/O should be an evident part of every approach towards user-friendly and accurate solvers.

For the derivatives notations, we imitate [Ort 70]. We use the concepts and notations of interval analysis as introduced in [Moo 69], [Ale 83] without any further explanations. For the screen of floating point numbers S and the corresponding operations $\boxplus$, $\boxminus$, $\boxdot$, $\boxslash$, see [Kul 81]. Interval operations in the interval space IS are denoted by $\Diamond$, $\Diamond$, $\Diamond$, $\Diamond$. A function name surrounded by a diamond $\Diamond$ means an enclosure of this function in IS.

2. Inclusion of Zeros

All known methods for an enclosure of zeros of a function $f: D \subseteq \mathbb{R}^n \to \mathbb{R}^n$ can be separated into two parts:

A-priori-methods: Starting with a (possibly large) interval $X^{(0)}$ containing a zero $\hat{x}$ of f, we compute successive iterates $X^{(k)}$ which also contain the same zero. The iterates have the property

$$X^{(0)} \supset X^{(1)} \supset X^{(2)} \supset \ldots \supset X^{(k)} \ni \hat{x}$$

A-posteriori-methods: We compute an approximation $\tilde{x}$ of a zero and apply a test procedure for a neighbourhood interval X of $\tilde{x}$ to verify a zero $\hat{x}$ within X.

Although bisection methods may be considered as a-priori methods, we restrict ourselves to methods which are based on the principle of interval analysis, contractive mappings and fixed-point theorems.

In the following, we start begin a broader explanation of both types of methods.

2.1 A-Priori-Methods

We shall often need the following property (cf. [Bau 87])

Definition 1: Let $X \in \mathbb{IR}^n$ and also let f be a function. We say that f is linearizable in X if a $L(f,X) \in \mathbb{IR}^{n \times n}$ exists that satisfies the relation
$$\forall x,y \in X : \exists \underset{\cdot}{L} \in L(f,X): f(x) - f(y) = \underset{\cdot}{L}(x - y) \quad (2)$$
We shall write briefly $L(X)$ instead of $L(f,X)$.

Definition 2: Let $f: D \subseteq \mathbb{R}^n \to \mathbb{R}^n$ be continuous, and for all intervals $Z \in \mathbb{IR}^{n \times n}$ with $Z \subset D$, linearizable.

Let $s: \mathbb{IR}^n \to \mathbb{R}^n$ be a selection function which selects from every $A \in \mathbb{R}^n$ an $s(A) \in A$. If for all intervals $Z \subset D$, an $H(Z) \in \mathbb{R}^{n \times n}$ exists satisfying

$$H(Z) \supseteq \{\dot{L}^{-1} \mid \dot{L} \in L(Z)\} , \tag{3}$$

then

$$N(Z,s) := s(Z) - H(Z) \cdot f(s(Z)) \tag{4}$$

is called the <u>interval</u> <u>Newton</u> <u>operator</u> of f.

If f is continuously differentiable, L can be replaced by f'. As a selction function, we often use the midpoint function

$$m(Z) := \frac{\inf(Z) + \sup(Z)}{2} \tag{5}$$

The fundamental theorem is the following

<u>Theorem 1:</u>

Let $f: D \in \mathbb{R}^n \to \mathbb{R}^n$ be continuous. Moreover, for every $Z \subset D$ with $Z \in \mathbb{R}^n$, f is linearizable and an $H(Z)$ as in (3) exists. Then, for the iteration

$$X^{(0)} := D,$$
$$X^{(k)} := N(X^{(k-1)},s) \cap X^{(k-1)} , \quad k = 1, 2, 3, \ldots$$

the following propositions hold:

(i)   If $X^{(0)}$ contains a zero $\hat{x}$ of f, then $\hat{x} \in X^{(k)}$, $k \geq 1$.

(ii)   If $N(X^{(k_0)}, s) \cap X^{(k)} = \phi$ for a $k_0 \geq 0$, then $X^{(0)}$ contains no zero.

(iii)   If s and H are continuous functions, every $\dot{H} \in H(D)$ is regular and $\hat{x} \in D$, then $\lim_{k \to \infty} X^{(k)} = \hat{x}$.

Proof: see [Bau 87], p.136 .

Proposition (iii) has only theoretical importance because computer representations of s and H are no longer continuous. If, for example, m(Z) is replaced by

$$\boxed{m}(Z) := \inf(Z) \boxplus (\sup(Z) \boxminus \inf(Z)) \boxslash 2,$$

then for any $Z \in IS^n$, $\boxed{m}(Z)$ lies in Z - but $\boxed{m}(Z)$ is not continuous!

Nevertheless, we have the following application on computers:

Corollary 1:
Suppose the conditions of theorem 1 hold for $f: D \in IS^n \to \mathbb{R}^n$. Then with

$$\langle\!\!\!\!\diamond\,N\,\diamond\!\!\!\!\rangle (Z, \boxed{s}) := \boxed{s}(Z) \diamond \langle\!\!\!\!\diamond\,H\,\diamond\!\!\!\!\rangle(Z) \diamond \langle\!\!\!\!\diamond\,f\,\diamond\!\!\!\!\rangle (\boxed{s}(Z))$$

where $\boxed{s}: IS^n \to S$ is a selection function in IS^n and $\langle\!\!\!\!\diamond\,H\,\diamond\!\!\!\!\rangle(Z) \in IS^{n \times n}$

satisfying

$$\langle\!\!\!\!\diamond\,H\,\diamond\!\!\!\!\rangle(Z) \supseteq H(Z) \supseteq \{L^{-1} | L \in L(Z)\}$$

for all $Z \in IS^n \cap PD$, the iteration

$$X^{(0)} := D$$

$$X^{(k)} := \langle\!\!\!\!\diamond\,N\,\diamond\!\!\!\!\rangle (X^{(k-1)}, \boxed{s}) \cap X^{(k-1)}, \quad k = 1, 2, 3, \ldots$$

in IS^n satisfies the following propositions:

(i)   If $X^{(0)}$ contains a zero $\hat{x}$ of f, then $\hat{x} \in X^{(k)}$, $k \geq 1$.

(ii)   If $\langle\!\!\!\!\diamond\,N\,\diamond\!\!\!\!\rangle (X^{(k_0)}, \boxed{s}) \cap X^{(k_0)} = \phi$ for $k_0 \geq 0$, then $X^{(0)}$

         contains no zero.

Of most importance for this kind of method, is a first enclosure X of a zero $\hat{x}$ which then can be improved more and more. This situation is similar to that of a bisection method where we have to find two different points with distinct function value signs. The only difference is that bisection methods allow multiple zeros and clusters of zero within the starting interval whereas theorem 1 and condition (3) imply the uniqueness of $\hat{x}$.

Note also that the inversion of a matrix with interval entries has to be performed. This can be done by the interval version of the Gauß-Jordan-Algorithm or by an iterative method described in [Rum 83].

2.2 A-Posteriori-Methods

We start with a very simple test of a set X (an interval) contains a zero. For this purpose, we state a theorem from [Ort 70]:

Lemma 1:   Suppose that g: $D \subseteq \mathbb{R}^n \to \mathbb{R}^n$ is contractive on a closed set $D_0 \subset D$ and $g(D_0) \subset D_0$. Then g has a unique fixed point in D_0.

Contractive means, of course, that there is a (global) positive constant α < 1 which satisfies a relation

$$\|g(x) - g(z)\| \leq \alpha \|x - z\|, \quad \forall\, x, z \in D_0 \quad .$$

With the notation $\rho(A)$ for the spectral radius of a matrix A we are now able to prove

Theorem 2:   Let $X \in \mathbb{IR}^n$ and f be a linearizable function on X. Suppose that a matrix $R \in \mathbb{R}^{n \times n}$ exists for which

$$\|I - R \cdot L(X)\| < 1 \qquad\qquad\qquad (6)$$

holds. Then f has a unique zero $\hat{x}$ in X.

Proof: Consider the function $g(x) := x - R \cdot f(x)$. Let $x, y \in X$. Since f is linearizable we have for a $L \in L(X)$

$$g(x) = x - R \, (f(y) + L(x-y) \,)$$
$$= x - Rf(y) + (I - R \cdot L)(x-y)$$
$$= g(y) + (I - R \cdot L)(x-y)$$

and from that

$$\|g(x) - g(y)\| \leq \|I - R \cdot L\| \, \|x-y\|$$
$$\leq \|I - R \cdot L(X)\| \cdot \|x-y\|$$

where $\alpha := \|I - R \cdot L(X)\|$ is a constant lower than 1. Therefore, g is contractive with a unique fixed point $\hat{x} \in X$. Since

$$\rho(I - R \cdot L) \leq \|I - R \cdot L(X)\| < 1$$

for every $L \in L(X)$ we have also that R and all L are nonsingular (consider the eigenvalues of $I - R \cdot L$). This leads to

$$\hat{x} = g(\hat{x}) = \hat{x} - R \cdot f(\hat{x}) \implies f(\hat{x}) = 0 \quad .$$

Since the fixed points of g and the zeros of f are equivalent, $\hat{x}$ is the unique zero of f.

Theorem 2 delivers a quite simple proof of existence and uniqueness of a zero which can be easily performed on computers. For L(X) we can take f'(X) if f is continuously differentiable. In this case the main task is to evaluate f'(X) and to compute a rather smooth inclusion of $I - R \cdot f'(X)$, with the hope that the norm of this term becames smaller then unity. This would be more likely if $R \approx f'(\tilde{x})^{-1}$ where $\tilde{x}$ is some value from X.

For the following, we have to state two further lemmas:

Lemma 2:   Let X be a nonempty, convex and compact subset of $\mathbb{R}^n$ and f:X → X continuous. Then f has at least one fixed point.

Proof: This is a simple consequence of Brower's fixed point theorem ([Heu 86]).

Lemma 3:   Let $X, Z \in \mathbb{IR}^n$, $B \in \mathbb{IR}^{n \times n}$. If $Z + B \cdot X \subsetneq X$, then for all $B \in B$ we have $\rho(B) < 1$ ($\subsetneq$ is defined componentwise and means "subset and not equality").

Proof: is rather technical and can be found in [Bau 88], p. 101.

We will use the operator $\underline{U}$ for the convex hull of two intervals, i.e. the smallest enclosing interval. Now we are able to state the following

Theorem 3:   Let $\tilde{x} \in \mathbb{R}^n$, and $X \in \mathbb{IR}^n$ and $f: \tilde{x} \underline{U} X \rightarrow \mathbb{R}^n$ a linearizable function. Suppose that

$$G(X) := \tilde{x} - R \cdot f(\tilde{x}) + (I - R \cdot L(\tilde{x}\underline{U}X)) \cdot (X - \tilde{x}) \subsetneq X \qquad (7)$$

for a given matrix $R \in \mathbb{R}^{n \times n}$ then R and all $L \in L(\tilde{x}\underline{U}X)$ are nonsingular and f has one and only one zero $\tilde{x} \in X$.

Proof: Consider the continuous function $g(x) := x - Rf(x)$ in X. Since f is linearizable we have an $L \in L(\tilde{x}\underline{U}X)$ with

$$g(x) = x - R(f(\tilde{x}) + L(x - \tilde{x}))$$
$$= \tilde{x} - R \cdot f(\tilde{x}) + (I - R \cdot L)(x - \tilde{x}) \in G(X)$$

Lemma 2 implies, therefore, the existence of a fixed point $\hat{x}$ of g in X. By Lemma 3 we find that, in addition, all $M \in I - R \cdot L(\tilde{x}\underline{U}X)$ are nonsingular. A simple consideration of the eigenvalues of all M shows that moreover R and all $L \in L(\tilde{x}\underline{U}X)$ are nonsingular. Therefore,

$$\hat{x} = g(\hat{x}) = \hat{x} - R \cdot f(\hat{x})$$

implies that $f(\hat{x}) = 0$. If $\hat{y}$ is another zero of f in $\hat{X}$, we have by (6) an $\underset{.}{\overline{L}} \in L(\tilde{x} \underline{\cup} X)$ satisfying

$$f(\hat{y}) = f(\hat{x}) + \underset{.}{\overline{L}}(\hat{x} - \hat{y})$$

Since $\underset{.}{\overline{L}}$ is nonsingular it follows that $\hat{x} = \hat{y}$.

Similar to theorem 2, we can take for $L(\tilde{x} \underline{\cup} X)$ an interval extension of the Jacobian f' of f. R can be chosen arbitrarily but we are well advised to define R as an approximation of $f'(\tilde{x})^{-1}$, so to make it more likely to satisfy (7) for an $X \in \mathbb{IR}^n$.

Theorem 3 is also applicable on computers. This proves the following

Corrollary 2:   Let $\tilde{x} \in S^n$, $Y \in IS^n$ and $f: \tilde{x} \underline{\cup} (\tilde{x} \diamond Y) \to \mathbb{R}^n$ a linearizable function.

Suppose that

$$\tilde{G}(Y) := - R \diamond \left\langle\!\!\diamond\!\!\!\left\langle f \right\rangle\!\!\diamond\!\!\right\rangle(\tilde{x}) \diamond \diamond (I - R \cdot L(\tilde{x} \underline{\cup} (\tilde{x} \diamond Y))) \diamond Y \underset{\neq}{\subsetneq} Y \qquad (9)$$

for a given matrix $R \in S^{n \times n}$, then R and all $\underset{.}{L} \in L(\tilde{x} \underline{\cup} (\tilde{x} \diamond Y))$ are nonsingular and f has one and only one zero $\hat{x} \in \tilde{x} \diamond Y$.

Proof: In difference to theorem 3 we have replaced X by $\tilde{x} \diamond Y$. Note also that theorem 3 remains true if $G(X)$ is replaced by any superset $H(X)$ and condition (7) by $H(X) \underset{\neq}{\subsetneq} X$.

Corollary 2 is actually a verification theorem for the error $\Delta x := \hat{x} - \tilde{x}$. Rewriting theorem 3 in this form has more technical reasons than mathematical ones.

For a somewhat surprising connection between theorem 2 and theorem 3 we first mention the simple interval arithmetic rules.

(a) For $A \in \mathbb{IR}$, $B = [-b,b] \in \mathbb{IR}$, $b \geq 0$ we have
$$A \cdot B = |A| \cdot b[-1,1].$$

(b) For $A = [-a,a]$, $B = [-b,b] \in \mathbb{IR}$, $a, b \geq 0$ we have
$$A + B = (a+b) \cdot [-1,1].$$

From this it follows that for any interval matrix $A \in \mathbb{IR}^{n \times n}$ and any interval vector $X \in \mathbb{IR}^n$ of the form $X = [-1,1] \cdot x$, $x \geq 0$ componentwise, the following equivalence holds

$$A \cdot X = [-1,1] \, |A| \cdot x,$$

where $|A|$ is the matrix of the absolute values of the components of A. If additionally all components x_i are equal and , say, identical to $\alpha > 0$, then

$$A \cdot X = [-1,1] \, \alpha \cdot \| \, |A| \, \|_\infty.$$

This means that

$$A \cdot X \subseteq X$$

is equivalent to

$$\| \, |A| \, \|_\infty \leq 1 \quad .$$

This fact shows that theorem 2 and theorem 3 are equivalent if applied to an enclosure Y of the error $x - \tilde{x}$ of the form $Y = \alpha[-1,1] \cdot (1,\ldots,1)^T$ with any positive constant α.

Is this a neglectable case? Not if we have a rather good approximation $\tilde{x}$ of a zero x already computed (probably stored in several succeeding registers). Then we are well advised to take a symmetric interval Y containing a zero as a final correction to be validated. If so, we can use theorem 2 rather then theorem 3 to check whether $\tilde{x} + Y$ contains the zero.

There is still an open question as to what happens if Y is only approximately a symmetric interval. Theoretically the assumption $G(Y) \subsetneq Y$ of theorem 3 could be valid but $\|I\text{-}R\cdot L(x+Y)\| > 1$ and theorem 2 can not be applied. Many practical examples published in [Rum 80] demonstrate that this theoretical case could actually happen but is rare. This leads us to the suggestion to always compute at first the comparably cheap norm of the operator $C := \Diamond(I\text{-}R\cdot\langle f'\rangle(X))$, and only if this fails, to start the iteration introduced in [Kau 82]:

$$
\begin{array}{l}
Y(0) := - R \odot \langle f' \rangle (\tilde{x}) \\[2mm]
k := 0 \; ; \\
\underline{\text{repeat}} \\[2mm]
\quad k := k{+}1 \; ; \quad Y(k{-}1) := Y(k{-}1) \circ \epsilon \; ; \\
\quad Y(k) := - R \odot \langle f' \rangle (\tilde{x}) \odot C \odot Y(k{-}1) \\[2mm]
\underline{\text{until}} \; Y(k) \subsetneq Y(k{-}1) \quad \underline{\text{or}} \quad k = k_{max};
\end{array}
$$

Here "∘" is componentwise the so-called ϵ-extension which is defined as

$$
\forall X \in \mathbb{IR}, \; \epsilon > 0 : \quad x \circ \epsilon := \begin{cases} X + \epsilon[-1,1]\cdot|X| & > 0 \\ [-\delta,\delta] \; , & |X| > 0 \end{cases}
$$

where $\delta > 0$ is the smallest positive computer representable number. As a rule of thumb the sum of ϵ and the contraction constant of the unerlaying operator must be lower than 1. It turns out that 0.1 is a good value for ϵ.

Another important way to check whether a set X contains a zero is the one below:

<u>Theorem 4:</u> Let $\tilde{x} \in \mathbb{R}^n$, $X \in \mathbb{IR}^n$ and $f : \tilde{x} \cup X \to \mathbb{R}^n$ a linearizable function. Suppose that

$$
G(X) := \tilde{x} - H(\tilde{x} \cup X) \cdot f(\tilde{x}) \subsetneq X \tag{10}
$$

where $H(\tilde{x} \underline{\cup} X) \in \mathbb{R}^{n \times n}$ and

$$H(\tilde{x} \underline{\cup} X) \supseteq \{\underline{L}^{-1} | \underline{L} \in L(\tilde{x} \underline{\cup} X)\} \tag{11}$$

then f has one and only one zero $\hat{x}$ in X.

Proof: For every $x \in X$ we have

$$f(x) = f(\tilde{x}) + \underline{L}(x) \cdot (x - \tilde{x}) \tag{12}$$

with an $\underline{L}(x) \in L(\tilde{x} \underline{\cup} X)$. $\underline{L}(x)$ is continuous and from (11) we see that $\underline{L}(x)$ is nonsingular. So

$$x - \underline{L}(x)^{-1} f(x) = \tilde{x} - \underline{L}(x)^{-1} \cdot f(x) =: g(x)$$

and $g(x) \in G(X)$. By (10) and lemma 1 we therefore have the existence of a fixed point $\hat{x}$ of g in X, i. e.

$$\hat{x} = \hat{x} - \underline{L}(\hat{x})^{-1} \cdot f(\hat{x})$$

which implies $f(\hat{x}) = 0$.

The proof of uniqueness is shown as in theorem 2.

We immediately state a corresponding computer arithmetic version of this theorem 4 as already done for theorems 2 and 3:

Corollary 3: Let $\tilde{x} \in \mathbb{R}^n$ and $f: \tilde{x} \underline{\cup} (\tilde{x} \diamondsuit Y) \rightarrow \mathbb{R}^n$ a linearizable function. Suppose that

$$\tilde{G}(Y) := - H (\tilde{x} \underline{\cup} (\tilde{x} \diamondsuit Y)) \diamondsuit f(\tilde{x}) \subsetneqq Y, \tag{13}$$

where $H (\tilde{x} \underline{\cup} (\tilde{x} + Y)) \in IS^{n \times n}$ and

$$H (\tilde{x} \underline{\cup} (\tilde{x} \diamondsuit Y)) \supseteq \{\underline{L}^{-1} | \underline{L} \in L(\tilde{x} \underline{\cup} (\tilde{x} \diamondsuit Y))\} \tag{14}$$

then f has one and only one zero $\hat{x}$ in $\tilde{x} \diamond Y$.

Again we have chosen the "correction form" for practical reasons.

2.3 Connection between a-priori and a-posteriori methods

It is obvious that there is a simple connection between the methods from sections 2.1 and 2.2. The a-priori method needs a first inclusion of a zero of f. The a-posteriori methods test a given set X to see if it contains a zero. For that we state the following

Theorem 4:   Let $\tilde{x} \in \mathbb{R}^n$, $X \in \mathbb{IR}^n$ and $f: \tilde{x} \cup X \to \mathbb{R}^n$ a linearizable function. Let $G(X)$ be defined as in (7) or (10). Then we have the following propositions:

(1)   If $G(X) \subsetneq X$, then f has a unique solution $\hat{x} \in X$ and for the iteration

$$X^{(0)} := X$$
$$X^{(k)} := G(X^{(k-1)}), \quad k = 1, 2, 3..$$

we have $X^{(k)} \subseteq X^{(k-1)}$, $k \geq 1$, and $\hat{x} \in X^{(k)}$, $k \geq 0$

(b)   If $G(X) \cap X = \phi$, then f has no zero in X.

Proof: (a)   From theorem 2 and 3 we have immediately for both types of G the existence and uniqueness of a zero $\hat{x} \in X$. By induction and the inclusion monotonicity, it is easily shown that $X^{(k)} \subseteq X^{(k-1)}$.

(b)   Suppose $f(\hat{x}) = 0$, $\hat{x} \in X$. Then for G from (7) we have the corresponding

$$g(\hat{x}) = \hat{x} - R \cdot f(\hat{x}) = \hat{x},$$

thus, g has a fixed point $\hat{x} \in X$. From this we have $\hat{x} = g(\hat{x}) \in G(x)$ and therefore $X \cap G(X) \neq \phi$ in contradiction to our assumption. The same holds for G from (10):

$$g(\hat{x}) = \hat{x} - \underset{\cdot}{L}(\hat{x}) \cdot f(\hat{x}) = \hat{x}$$

for a nonsingular $\underset{\cdot}{L}(\hat{x}) \in \mathbb{R}^{n \times n}$.

A corresponding transformation to computer systems as in the corollaries above is straightforward and is omitted here.

3. Numerical problems with traditional methods

As shown in the last chapter the theory of validating zeros of a function does not replace "classical" methods of root finding but is a supplement. Theorems 3 and 4 are based on the fact that there is already a sufficient good approximation $\tilde{x}$. Consequently, we must consider methods which deliver such an approximation and examine what problems arise [Ort 70].

Traditional methods are e.g.
- Newton's and Newton-like methods
- minimization methods such as decent methods, Gauss-Newton and so forth
- Specialized methods for certain pre-structured problems like nonlinear Gauss-Seidel iteration, Newton SOR and s.o.

All mentioned methods have something in common which needs our attention. These points are discussed by means of the following example of Newton's method:

---

1. Choose an $x^{(0)}$

2. Iterate
   $$x^{(k+1)} := x^{(k)} - f'(x^{(k)})^{-1} \cdot f(x^{(k)}),$$
   $$k = 0, 1, 2, \ldots$$

Note that for each k ≥ 0 an iteration step is actually performed via

---

A(k) := f'(x(k))
b(k) := - f(x(k))
Solve the system A(k) · Δ(k) = b(k)
x(k+1) := x(k) + Δ(k)

This shows us the typical difficulties:

(a) <u>Evaluation of the function f:</u>

Of course, each method must compute values of f to check how near or far we are from zero. The function itself consists of many operations such as +, -, ·, / or calls of elementary functions like sin, cos, and so on. Some of the intermediate results must cancel each other in the near of a zero which tends to be a numerical problem as shown in many publications (cf. [Kah 71], [Böh 83], [Fis 88].). Note also, that theorems 3 and 4 need sharp inclusions of f at the point $\tilde{x}$.

(b) <u>Solution of linear systems of equations:</u>

Most methods work with local linearizations of the function f which lead to linear systems of equations, sometimes of special shape. Although a wrong computed direction for the next iterate is not as important as a wrong function value, it may significantly influence the convergence of the iteration. If, for example, in Newton's method f' is evaluated in the neighborhood of singularity the corresponding matrix A is nearly singular. Therefore, the problem of solving A·Δ=b may be ill-conditioned.

(c) <u>Evaluation of partial derivatives:</u>

Some methods also need information about the partial derivatives of f at certain points. Here we have two different methods:

- Approximation of $\partial_j f_i(x)$ by a term like

$$\frac{f_i(x_1^{(1)}, x_2^{(1)}, \ldots, x_j^{(2)}, \ldots, x_n^{(1)}) - f_i(x^{(1)})}{x_j^{(2)} - x_j^{(1)}}$$

for different points $x^{(1)}$ and $x^{(2)}$

- Automatic differentiation techniques which supply the possibility of computing an analytical derivative of f during the "normal" evaluation of f (see [Ral 81]).

While the first method has the disadvantage of delivering bad approximations due to cancellation, the second may need, in general, more storage and computing time. Again, this point is not as important as the function value itself but it should be treated with care.

In the following, we give a broader explanation of these points and ways in which to deal with them.

3.1 Evaluation of function values

We will not expound here the difficulties which arise on computers when a function $f(x)$ is replaced by an approximation $\boxed{f}(x)$. The latter one is created by a replacement of all real operations ∘ in f by corresponding computer approximations ⊡. These difficulties are already well discussed in [Wil 63] and [Kah 71] and many other publications.

We restrict ourself to reporting some ways around these problems which can be completely done by the computer itself. The main remedy is the use of interval arithmetic instead of approximation arithmetic. So we have for each operation ∘ the computer interval version ◈ with the well-known property

$$\forall X, Y \in IS; \ \forall x \in X, y \in Y : x \circ y \in X \diamond Y \tag{15}$$

where S denotes the used screen of floating point numbers and IS the corresponding interval space. This leads to the fundamental relation

$$f(x) \in \left\langle\!\!\left\langle f \right\rangle\!\!\right\rangle (Y) \qquad \forall x \in X \in IS, \tag{16}$$

if $\langle\!\!\langle f \rangle\!\!\rangle$ is produced from f by using the mentioned interval operations [Moo 69] and [Ale 83].

But (16) provides only inclusion and we have, in general, no foreknowledge as to how large the diameter $d(\langle\!\!\langle f \rangle\!\!\rangle(x))$ tends to be. Nevertheless, there is a strong relation between the mantissa length of the used floating point screen and the diamter of the resulting interval $\langle\!\!\langle f \rangle\!\!\rangle(x)$. Therefore, it is clear to expand the mantissa length to reach as much accuracy as we need.

A simple strategy might be to have a first inclusion using a mantissa length t_1, resulting in $\langle\!\!\langle f \rangle\!\!\rangle_{t_1}(X)$. If $d(\langle\!\!\langle f \rangle\!\!\rangle_{t_1}(x))$ is too large we increase t_1 to t_2, recompute $\langle\!\!\langle f \rangle\!\!\rangle_{t_2}(x)$, check $d(\langle\!\!\langle f \rangle\!\!\rangle_{t_2}(x))$ and so on. Finally we reach a $\langle\!\!\langle f \rangle\!\!\rangle_{t_n}$, $n \geq 1$, with a sufficiently tight diameter.

To do this, we need something like a multiple precision arithmetic. Such an arithmetic can be provided using techniques from [Knu 81] combined with those from [Kul 81] which are based purely on integer arithmetic ([Bre 80] and [Bre 81]). Another approach would be to use floating point arithmetic based on similar properties as IEEE 754, a standard for binary floating point arithmetic [ANS 85]. Higher precision can be achieved by using methods from [Dek 71] and [Lin 81] or by applying an exact scalar product as introduced in [Kul 81] and extensively discussed and examined within the DIAMOND project.

The latter approach could also result in a technique developed during DIAMOND Task 2a [Fis 88]. This method does not provide multiple precision operators explicitly, but automates the process of recomputation in an iterative way and is based on an earlier approach in [Böh 82]. Certain advantages and disadvantages of this method are discussed at the given references and are omitted here.

All procedures can also be extended from pure algebraic functions to functions where transcendental expressions like sin, cos, ... appear. Recent research results ([Bra 88] and [Krä 88]) are available which provide the possibility of evaluating those elementary functions with arbitrary accuracy, even with intervals as arguments.

3.2 Solving linear systems

There is also a DIAMOND task (T3-1), which deals with systems of linear equations. It consists of an application of results from [Rum 80] to complex coefficient matrices. Those techniques deliver least bit accuracy for the solution if there is any. The mathematical background is the same as stated here in chapter 2. For high accuracy, the exact scalar product is used.

Of course, all considerations of the previous section 3.1 are valid if applied to any direct method known for linear systems (Gaussian elimination, LU-decomposition, ...). Therefore, multiple precision arithmetic could be used also to achieve better approximations for the solution.

Note that a linear system solver is only needed in the approximation step where we are not interested in inclusions of intermediate iterates. But inclusion is the only way to come to a guaranteed assertion about the accuracy of intermediate iterates.

3.3 Computation of partial derivatives

Methods for automatic differentiation have been known for a long time. Nevertheless, it is not possible to say where it was stated first. An extensive presentation of automatic differentiation is given in [Ral 81]. We give here an exemplary description of this technique.

Suppose the function $f: D \subseteq \mathbb{R}^n \to \mathbb{R}$ is defined by the following recursion:

Let $x \in D$ be fixed, then

$$
\begin{aligned}
z_1 &:= \phi_1(x) \\
z_2 &:= \phi_2(x, z_1) \\
&\vdots \\
z_m &:= \phi_m(x, z_1, \ldots, z_m) = f(x)
\end{aligned}
\tag{17}
$$

The $z_i \in \mathbb{R}$, $i = 1, \ldots, m$, are to be considered as intermediate results while computing $f(x)$. The function $\phi_i: D \times \mathbb{R}^{i-1} \to \mathbb{R}$ represents a certain operation on components of x and some former intermediate results. If $\frac{\partial}{\partial x_j} f(x)$ has to be computed for some $j \in \{1, \ldots, n\}$ we find recursively backwards that

$$
\begin{aligned}
\frac{\partial z_1}{\partial x_j} &= \frac{\partial \phi_1}{\partial x_j}(x) \\
\frac{\partial z_2}{\partial x_j} &= \frac{\partial \phi_2}{\partial x_j}(x, z_1) + \frac{\partial \phi_2}{\partial z_1}(x, z_1) \\
&\vdots \\
\frac{\partial z_m}{\partial x_j} &= \frac{\partial \phi_m}{\partial x_j}(x, z_1, \ldots, z_{i-1}) + \sum_{k=1}^{i-1} \frac{\partial \phi_m}{\partial z_k}(x, z_1, \ldots, z_{i-1}) \cdot \frac{\partial z_k}{\partial x_j}
\end{aligned}
\tag{18}
$$

Therefore, we have also a recursion for computing $\frac{\partial f}{\partial x_j} = \frac{\partial z_m}{\partial x_j}$ with intermediate results $\frac{\partial z_1}{\partial x_j}$, $i = 1, \ldots, m$.

In practice, ϕ_i is usually an unary operation on either a x_k or a z_l, $1 \leq l \leq i-1$, or ϕ_i is a binary operation between two x_k's, two z_l's or a x_k and z_l. This fact leads to very simple formulas (18) as the following example shows:

<u>Example:</u>

Let

$$f(x_1, x_2) := \frac{x_1^6 x_2 + 3 x_1^4}{7x_2} \quad .$$

We want to compute f and $\frac{\partial f}{\partial x_1}$:

$z_1 := x_1$ $\qquad\qquad$ $\frac{\partial z_1}{\partial x_1} = 1$

$z_2 := z_1^6$ $\qquad\qquad$ $\frac{\partial z_2}{\partial x_1} = 6 \cdot z_1^5 \cdot \frac{\partial z_1}{\partial x_j}$

$z_3 := x_2$ $\qquad\qquad$ $\frac{\partial z_3}{\partial x_1} = 0$

$z_y := z_2 \cdot z_3$ $\qquad\qquad$ $\frac{\partial z_4}{\partial x_1} = z_2 \cdot \frac{\partial z_3}{\partial x_1} + z_3 \cdot \frac{\partial z_2}{\partial x_1}$

$z_5 := 3$ $\qquad\qquad$ $\frac{\partial z_5}{\partial x_1} = 0$

$z_6 := x_1$ $\qquad\qquad$ $\frac{\partial z_6}{\partial x_1} = 1$

$z_7 := z_5 \cdot z_6$ $\qquad\qquad$ $\frac{\partial z_7}{\partial x_1} = z_5 \cdot \frac{\partial z_6}{\partial x_1} + z_6 \cdot \frac{\partial z_5}{\partial x_1}$

$z_8 := z_4 + z_7$ $\qquad\qquad$ $\frac{\partial z_8}{\partial x_1} = \frac{\partial z_4}{\partial x_1} + \frac{\partial z_5}{\partial x_1}$

$z_9 := 7$ $\qquad\qquad$ $\frac{\partial z_9}{\partial x_1} = 0$

$z_{10} := x_2$ $\qquad\qquad$ $\frac{\partial z_{10}}{\partial x_1} = 0$

$z_{11} := z_9 \cdot 7_{10}$ $\qquad\qquad$ $\frac{\partial z_{11}}{\partial x_1} = z_9 \frac{\partial z_{10}}{\partial x_1} + z_{10} \frac{\partial z_9}{\partial x_1}$

$z_{12} := z_8/z_{11}$ $\qquad\qquad$ $\frac{\partial z_{12}}{\partial x_1} = (z_{11} \frac{\partial z_8}{\partial x_1} - z_8 \frac{\partial z_{11}}{\partial x_1})/z_{11}^2$

We conclude this section with two remarks for the practical application:

- The explicit storage of all z_i and $\frac{\partial z_i}{\partial x_j}$ can be avoided if well-known stack techniques are used. Then, only those intermediate results have to be stored which are needed for the computation process to follow.

- In the computation formulas for the derivatives, we need also the values of the function itself. If a stack technique is used the intermediate result of a certain derivative has to be updated before the corresponding intermediate result of the function value itself.

- This concept can be easily extended to higher derivatives.

4. Improvement of theoretical behaviour of traditional methods

Most iterative methods are not convergent for an arbitrary initial guess. They have only local convergence. The fixed points of those iterations have often very curious areas of attraction as can been seen by certain fractals or Mandelbrot's figures.

What happens if we chose an initial guess outside the (theoretical) domain of local convergence?

4.1 Starting phase of Newton's method

Let us consider again Newton's method. The starting value $x^{(0)}$ is improved by adding a term $\Delta^{(0)} := -f'(x^{(0)})^{-1} \cdot f(x^{(0)})$. This should us lead to a new value $x^{(1)} := x^{(0)} + \Delta^{(0)}$ which should be closer to an assumed zero $\hat{x}$ of f. What does "closer" mean?

A measure for this could be the norm of f, for example, if

$$\|f(x^{(k+1)})\| \leq \|f(x^{(k)})\| \tag{19}$$

for successive iterates $x^{(k)}$ and $x^{(k+1)}$. But this must not hold even if the sequence $x^{(k)}$ converges to $\hat{x}$. Therefore, a simple modification of Newton's method is

$$x^{(k+1)} := x^{(k)} - \mu_k \cdot f'(x^{(k)})^{-1} \cdot f(x^{(k)}), \tag{20}$$

where the factors μ_k are chosen to ensure (19) (see [Ort 70]).

An algorithm could look as follows:

---

1. Choose $x^{(0)} \in \mathbb{R}^n$; k:=0

2. repeat

    k := k+1; j := -1;

    $y^{(k)} := f'(x^{(k-1)})^{-1} \cdot f(x^{(k-1)})$ {Newton direction}

    repeat

        j := j+1;

        $x^{(k,j)} := x^{(k-1)} - 2^{-j} \cdot y^{(k)}$

    until $\|f(x^{(k,j)})\| \leq \omega \cdot \|f(x^{(k-1)})\|$;

    $x^{(k)} := x^{(k,j)}$

  until $\|x^{(k)} - x^{(k-1)}\| \leq \epsilon \cdot \|x^{(k)}\|$;

Here successive bisection is chosen to determine a μ_k as in (20). The constants ω and ϵ could be user defined. We set $\omega = 0,5$. A security upper bound for the counter of inner and outer iterations should also be provided.

4.2 Continuation methods

A very common way to overcome the difficulties in finding an appropriate initial guess is the use of a continuation method (cf. [Ort 70]). Instead of the original problem $f(x) = 0$, one considers the problem

$$g(x,t) = 0 \tag{21}$$

where t varies within [0,1] and g is a function with the property

$$g(x^{(0)},0) = 0, \; g(x,1) = f(x) \tag{22}$$

with a given vector $x^{(0)}$. As an example we can take

$$g(x,t) := f(x) + (t-1) \cdot f(x^{(0)}) \qquad (23)$$

and try to solve (21). g surely satisfies (22).

It is well-known, that (21) describes a curve $x(t)$ in $\mathbb{R}^n$ when t runs from 0 to 1. Moreover, with (23) $x(t)$ satisfies the differential equation

$$\dot{x} = -f'(x)^{-1} \cdot f(x^{(0)}), \quad x(0) = x^{(0)} \qquad (24)$$

The principle of continuation methods is the following: starting with the known solution of $g(x,0) = 0$ we try to solve $g(x,t) = 0$ with a sufficiently small value of t. If this is done we increase t slightly, solve (21) again and so on. Algorithmically this is done by intersecting the interval $[0,1]$ into certain subintervals $0 < t_1 < t_2 < \ldots < t_m = 1$ with $m \in \mathbb{N}$ and successively solving $g(x,t_i) = 0$, $i = 1, 2, \ldots, m$.

As a starting value for the iteration in each step we can choose the solution for the last t-value. We can also use the relation (24) and perform an Euler-step from $x(t_{i-1})$ to $x(t_i)$ by

$$x(t_i) \approx x(t_{i-1}) - (t_i - t_{i-1}) \cdot f'(x(t_{i-1}))^{-1} \cdot f(x^{(0)}).$$

This was done in the following algorithm for a continuation method using the function (23):

---

1. Choose value $x^{(0)}$;

   i = 0

2. Perform one Euler-step

   $y := x^{(i)} - (t_{i+1} - t_i) \cdot f'(x^{(i)})^{-1} \cdot f(x^{(0)})$

3. Solve iteratively $g(x^{(i)}, t_i) = 0$ with initial guess y

4. i := i + 1;

   if $i \leq m$ goto 2.

---

5. Condition of a System of Nonlinear Equations

A first approach to define a condition number for a system of nonlinear equations $f(x) = 0$ is to define it as the condition number of the corresponding Jacobian f'. Obviously, this definition depends on the value x at which the Jacobian is computed. It becomes independent if we consider the Jacobian only at the (unknown) zeros $\hat{x}$ of f, which gives us a different condition number for each zero.

There is a simple objection for this first ad hoc approach: All one-dimensional problems seem to be well-conditioned since the defined condition number is equal to 1.

In [Woz 77] we find a more general strategy to define what "condition" means.

This strategy makes use of the fact that every nonlinear system depends in practice on some parameters, i.e.

$$f(x,p) = 0 \tag{20}$$

where $f: D \times P \to \mathbb{R}^n$, $D \subseteq \mathbb{R}^n$, $P \subseteq \mathbb{R}^m$. For instance, if $n = 1$ and f is a polynomial, then p may be considered as the vector of all coefficients of this polynomial. In most applications p is a vector of physical data, not seldom afflicted with tolerances. Thus, the term "condition" should describe how a change in the data vector disturbs the solution $\hat{x} = \hat{x}(p)$.

[Woz 77] uses the definition

$$\text{cond}(f,p) := \|\partial_x f(\hat{x},p)^{-1} \cdot \partial_p f(\hat{x},p)\| \cdot \frac{\|p\|}{\|\hat{x}\|} \tag{21}$$

where the norms $\|\cdot\|$ should be chosen very carefully. In the case of a linear system, (21) is equivalent to the well-known condition number of the system matrix where the used norm is the Schur norm. For polynomials $f(x) = \sum_{i=0}^{n} p_i x^i$ we have the relation

$$\text{cond } (f,p) = \sqrt{\sum_{i=0}^{n} |\hat{x}|^{2i-2}} \cdot \frac{\| p \|_2}{f'(\hat{x})} \qquad .$$

With that, we have an estimation of the error $\tilde{x} - \hat{x}$ for some approximation $\tilde{x}$

$$\frac{\| \tilde{x} - \hat{x} \|}{\| \hat{x} \|} \leq C \cdot b^{-t} \cdot \text{cond}(f,p) + O(b^{-2t}), \qquad (22)$$

where b is the base of the computer in use and t the mantissa length.

Thus, cond(f,p) gives us a "feeling" for the sensitiveness of the problem. No matter how good our computed inclusion of $\hat{x}$ ought to be, this quantity should be displayed also. It prevents the user of a naive use of a high accuracy result since when cond(f,p) is large, a slight disturbance of p would change $\hat{x}(p)$ significantly.

We remark that in our prototype we have always used the norm $\| \cdot \|_2$.

6. Implementation aspects

As part of the DIAMOND project we also implement a prototype of a solver for systems of nonlinear equations in PASCAL SC on an ATARI ST called NLSS. It should provide an interactive facility to help the user to find a zero and to compute sharp bounds. As a man-mashine interface we choose the same surface as it was done for the Online Training Component of IBM'S program package ACRITH.

The whole algorithm is separated into two parts: an approximation step and an inclusion step. These two steps are repeated with increasing accuracy if

- the approximating iteration diverges or
- no inclusion could be achieved or
- the inclusion does not suffice.

The method for approximation is Newton's method since the computation of the partial derivatives must be possible even for the inclusion method presented in chapter 2, if we take $f'(x)$ for $L(X)$. The algorithm includes the technique for the starting phase of Newton's method outlined in chapter 4.

If all else fails, a continuation method could be started of the form

$$g(x,t) := f(x) + (t-1) \cdot f(x^{(0)}) = 0$$

outlined in chapter 4. See that chapter for further explanations.

We summarize here the algorithmical parts of the preceeding chapters into a whole algorithm. At certain steps further points have been added as, for example, some Gauß-Seidel-iterations. This would be helpful in cases where the system has triangular form.

All operations are floating point operations.

1. Choose $x^{(0)} \in IS^n$; $k := \phi$; Starting-phase := true ;

   $q := 0$;

2. $k := k + 1$;

3. <u>if</u> $k > 1$, <u>go to</u> 4.

   {Gauß-Seidel-Iteration}

   <u>for</u> $i := 1$ <u>to</u> n

        $x^{(0,0)} := x^{(0)}$; $j := 0$;

        <u>repeat</u>

            $j := j + 1$; $z := x_i^{(0,j-1)}$;

            $d := (\frac{\partial}{\partial x_i} f_i(x^{(0,j-1)}))^{-1} \cdot f_i(x^{(0,j-1)})$

            $x^{(0,j)} := x^{(0,j-1)} - d \cdot e^{(i)}$

        <u>until</u> $|d| < \epsilon|z|$ <u>or</u> $j = jmax_1$

   <u>end</u> <u>for</u>;

4. $q := q + 1$

 $b := f(x^{(k-1)})$; $A := f'(x^{(k-1)})$

 <u>Solve</u> $A \cdot d^{(k)} = b$;

 <u>If</u> no error occurs, <u>go</u> <u>to</u> 5.

 Vary $x^{(k-1)}$ and <u>go</u> <u>to</u> 4.

5. if not starting-phase, goto 6.

 $d^{(k,0)} := d^{(k)}$; $j := \phi$;

 <u>repeat</u>

 $j := j + 1$; $d^{(k,j)} := 2^{j-1} \cdot d^{(k,j-1)}$;

 <u>until</u> $\|f(x^{(k-1)} - d^{(k,j)})\|_\infty < \mu \| f(x^{(k-1)})\|_\infty$ <u>or</u> $j = jmax_2$;

 $d^{(k)} := d^{(k,j)}$;

6. $x^{(k)} := x^{(k-1)} - d^{(k)}$; {New iterate}

 Starting-phase $:= \max\limits_{i=1,\cdots,n} |\dfrac{d_i^{(k)}}{x_i^{(k)}}| > \delta$;

 $x_i^{(k)} \neq 0$

7. If $|d_i^{(k)}| \le \epsilon \cdot |x_i^{(k)}|$ for all $i = 1,..,n$, goto 8.

 <u>If</u> $k = k_{max_1}$, <u>stop</u> with error.

 <u>Go</u> <u>to</u> 2.

8. Approximation $\tilde{x} := x^{(k)}$;

9. $U := \Diamond(f(\tilde{x}))$; {Residuum}

 Compute approximation R of $f(\tilde{x})^{-1}$;

 $Z := - R \Diamondtimes U$;

 $k := 0$; $Y^{(0)} := Z$;

10. <u>repeat</u>

 $k := k + 1$; $Y^{(k-1)} := Y^{(k-1)} \circ \gamma$;

 $Y^{(k)} := Z \Diamondtimes \Diamond(I - R \cdot f(\tilde{x} \Diamondtimes Y^{(k-1)} \underline{\cup} 0)) \Diamondtimes Y^{(k-1)}$

 <u>until</u> $Y^{(k)} \mathrel{\underset{\neq}{\subset}} Y^{(k-1)}$ <u>or</u> $k = k_{max_2}$;

11. <u>If</u> not $Y^{(k)} \mathrel{\underset{\neq}{\subset}} Y^{(k-1)}$, <u>stop</u> with error.

12. Solution lies in $X := \tilde{x} \Diamondtimes Y^{(k)}$

Some remarks should be made:

(a) In step 3 we perform some Gauß-Seidel-iterations. Note that this is only convergent on certain additional conditions on f (refer [Ort 70] for further explanations). One of these conditions is that $f(x) = 0$ represents a lower triagular system, i.e. $f_i(x) = f_i(x_1, \ldots, x_i)$.
The $e^{(i)}$ used in this step are the i-th unit vector (all components vanish except of the i-th which is 1). Note also, that step 3 is only performed for $k = 1$.

(b) If in step 4 the linear system seems to be singular, some variations of the starting value should be tried. We have used the form

$$x_i^{(k)} := x_i^{(k)} \cdot \sin(q+i) \cdot \beta$$

where β is a positive factor (see table below).

(c) Step 5 performs the shortening of the Newton direction as explained in chapter 4. This is only done during the starting phase of the iteration which is indicated by checking the condition in step 6.

(d) In step 10, we used the so called γ-extension of an interval vector which supports a rapid inclusion and which was explained in section 2.2.

(e) If one of the three error exits have to be used, either a continuation method or all computations with higher precision can be performed.

(f) The following table displays all parameters used in the algorithm, their meaning, and their default values at the start of the program:

name	meaning	default value
ϵ	Relative error bound (difference between two successive iterates)	10^{-10}
k_{max_1}	Maximum number of iterates in approximation	15
k_{max_2}	Maximum number of iterates in inclusion iteration	10
q_{max}	Maximum number of variations of the starting value	2
μ	Factor by which the norm of two successive iterates should decrease	$\frac{1}{2}$
j_{max_1}	Maximum number of Gauß-Seidel-steps	0
j_{max_2}	Maximum number of bisection steps in step 5.	10
δ	Maximum quotient between $|d_i^{(k)}|$ and $|x_i^{(k)}|$ at which convergence is assumed	10^{-2}
γ	Inflation factor	0.1

References

[Ale 83] Alefeld, G.; Herzberger, J.: Introduction to Interval Computation Academic Press, New York (1983)

[Bau 87] Bauch, H. et al.: Intervallmathematik. BSB B.G. Teubner Verlagsgesellschaft, Leibzig (1987)

[Böh 83] Böhm, H.: Berechnung von Polynomnullstellen und Auswertung arithmetischer Ausdrücke mit garantierter maximaler Genauigkeit. Dissertation, Universität Karlsruhe (1983)

[Böh 87] Böhm, H.; Rump, S.M.; Schumacher, G.: E-Methods for Nonlinear Problems, in [Kau 87], S. 59-80 (1987)

[Bra 88] Braune, K.: Standard Functions of Real and Complex Point and Interval Arguments with Dynamic Accuracy. Computing Suppl. 6, p 159-184 (1988)

[Bre 80] Brent, R.P.; Hooper, J.A.; Yohe, J. M.: An Augment Interface for Brent's Multiple-Precision Arithmetic Package. ACM Trans. Math. Software , S. 146-149(1980)

[Bre 81] Brent, R.P.: MP User's Guide (Fourth Edition). Technical Report TR-CS-81-08, Department of Computer Science, Australian National University, Canberra (1981)

[Dek 81] Dekker, T.J.: A Floating-Point Technique for Extending the Available Precision. Numer. Math. 18, S. 224-242 (1971)

[Fis 88] Fischer, H.C.; Schumacher, G.; Haggenmüller, R.: Evaluation of Arithmetic Expressions with Guaranteed High Accuracy. Computing Suppl. 6, p 149-158 (1988)

[Heu 86] Heuser, H.: Lehrbuch der Analysis, Band II, 3. Auflage. Teubner Verlag, Stuttgart (1986)

[Kah 71] Kahan, W.: A survey of error analysis, Proc. IFIP Congress 2, p. 1214-1239 (1971)

[Kau 82] Kaucher, E.; Rump, S.M.: E-Methods for Fixed-Point Equations f(x) = x. Computing 28, S. 31-42 (1982)

[Kau 87] Kaucher, E.; Kulisch, U.; Ullrich, Ch. (Hrsg.): Computer Arithmetic - Scientific Computation and Programming Languages. Teubner Verlag, Stuttgart (1987)

[Knu 81] Knuth, D.E.: The Art of Computer Programming, Volume 2 / Seminumerical Algorithms, 2. Auflage. Addison-Wesley, Reading (1981)

[Krä 88] Krämer, W.: Inverse Standard Functions for Real and Complex Point and Interval Arguments with Dynamic Accuracy. Computing Suppl. 6, p 185-212 (1988)

[Kul 81] Kulisch, U.; Miranker, W.: Computer Arithmetic in Theory and Practice. Academic Press, New York (1981)

[Kul 83] Kulisch, U.; Miranker, W.L.: A New Approach to Scientific Computation. Academic Press, New York (1983)

[Kul 87] Kulisch, U. (Hrsg.): PASCAL-SC: A PASCAL Extension for Scientific Computation, Information Manual and Floppy Disks, Version ATARI ST. Teubner Verlag, Stuttgart (1987)

[Lin 81] Linnainmaa, S.: Software for Double-Precision Floating-Point Computations, ACM Transactions on Mathematical Software 7, Nr. 3, S. 272-283 (1981)

[Ort 70] Ortega, J.M.; Rheinboldt, W.C.: Iterative Solution of Nonlinear Equations in Several Variables. Academic Press, Orlando (1970)

[Ral 81] Rall, L.B.: Automatic Differentiation: Techniques and Applications. Lecture Notes in Computer Science 120, Springer, Berlin (1981)

[Rum 80] Rump, S.M.: Kleine Fehlerschranken bei Matrixproblemen, Dissertation, Universität Karlsruhe (1980)

[Rum 83] Rump, S.M.: Solving Algebraic Problems with High Accuracy, in [Kul 83] (1983)

[Wil 63] Wilkinson, J.H.: Rounding Errors in Algebraic Processes. Her Majesty's Stationery Office, London (1963)

[Woz 77] Wozniakowski, H.: Numerical stability for solving Nonlinear Equations. Numer. Math. <u>27</u>, p 373-390 (1977)

[ACR 86] IBM High Accuracy Arithmetic Subroutine Library: Program Description and User's Guide. IBM Corporation (1986)

[ANS 85] ANSI/IEEE Standard 754-1985, Standard for Binary Floating-Point Arithmetic, New York (1985)